Fachwörterbuch Physik

Fachwörterbuch Physik

Autor: Dr. Matthias Heidrich, Kantonsschule Wil

Inhaltliche Beratung: Dr. Daniel Schläpfer, Kantonsschule Wil, Lehrbeauftragter der Universität Zürich

Sprachliche Beratung: Catherine Sandwell, BA, University of London, Kantonsschule Wil

Gestaltung: Knut Dewald, Stockum-Püschen

Matthias Heidrich, Dr. rer. nat., hat an der Universität Heidelberg Physik und Mathematik studiert. Nach langjähriger Forschungstätigkeit am CERN in Genf hat er zunächst in der Versicherungsmathematik bei Swisslife in Zürich gearbeitet, um dann Lehrer für Physik und Mathematik bei Minerva (AKAD-Gruppe) in Zürich zu werden. Seit 2004 ist er Lehrer für Physik und Astronomie in deutscher und englischer Sprache an der Kantonsschule Wil.

Daniel Schläpfer, Dr. sc. nat., hat an der Universität Zürich Geographie und Physik studiert. Neben seiner heutigen Forschungstätigkeit in Zusammenarbeit mit verschiedenen internationalen Universitäten, Instituten und Firmen ist er Lehrbeauftragter der Universität Zürich. Zudem unterrichtet auch er seit 2002 an der Kantonsschule Wil Physik in Deutsch und Englisch.

Catherine Sandwell hat an der University of London Deutsch und Englisch studiert und ist seit 2011 Englischassistentin an der Kantonsschule Wil.

Impressum:

© 2012 Matthias Heidrich
Herstellung und Verlag: Books on Demand GmbH, Norderstedt
ISBN: 978-3-8482-0918-7
Bibliografische Information der Deutschen Nationalbibliothek

Inhalt

Inhalt 5

Vorwort 7

Benutzungshinweise 8

Wortverzeichnis Englisch - Deutsch 9

Wortverzeichnis Deutsch - Englisch 35

Themenspezifische Wortverzeichnisse 61

    Methods / Methoden 62

    Kinematics / Kinematik 66

    Dynamics / Dynamik 68

    Gravitation / Gravitation 72

    Work, Power, Energy / Arbeit, Energie, Leistung 75

    Hydrostatics / Hydrostatik 76

    Heat / Wärme 79

    Electrostatics / Elektrostatik 84

    Electrodynamics / Elektrodynamik 87

    Magnetism and Electric Motor / Magnetismus und Elektromotor 92

    Induction and Generator / Induktion und Generator 95

    Oscillations / Schwingungen 96

    Waves / Wellen 98

    Acoustics / Akustik 101

    Radioactivity / Radioaktivität 103

    Optics / Optik 106

    Relativity / Relativität 112

    Quantum Physics / Quantenphysik 113

    Physical Constants / Physikalische Konstanten 114

    SI-base Quantities / SI-Basisgrößen 114

    Other physical Quantities / Andere physikalische Größen 114

    SI-base Units / SI-Grundeinheiten 115

    Derived SI-Units / Abgeleitete SI-Einheiten 115

- Non-SI-Units / Nicht-SI-Einheiten 116
- Technical Terms / Technische Begriffe 118
- Chemistry / Chemie 124
- Mathematics / Mathematik 129

Vorwort

Als Student eines naturwissenschaftlichen Fachs an der Universität oder Hochschule wie auch als Schüler im Bilingualunterricht sieht man sich zunehmend mit der Anforderung konfrontiert, englischsprachige Physikbücher oder Arbeitsmaterialien lesen und verstehen zu müssen. Ebenso will man auch selbst Arbeiten und Veröffentlichungen in englischer Sprache verfassen.

Sucht man in den herkömmlichen Wörterbüchern, sei es in Buch- oder elektronischer Form, nach den Übersetzungen von physikalischen Fachbegriffen, so wird man kaum fündig oder gar in die Irre geführt.

Dieses Fachwörterbuch hingegen bietet die im mathematisch-naturwissenschaftlichen Zusammenhang korrekten Übersetzungen von ca. 2200 grundlegenden Begriffen aus allen Gebieten der Physik einschließlich Begriffen aus der Technik, der Chemie und der Mathematik vom Deutschen ins Englische und umgekehrt.

Neben der alphabetischen Aufstellung dieser Begriffe wie in einem herkömmlichen Fachwörterbuch enthält dieses Werk zudem die Gruppierungen dieser Wörter nach 26 Themen wie Radioaktivität, Hydrostatik, Elektrodynamik oder Mathematik. Diese themenspezifischen Wortverzeichnisse eignen sich zum Beispiel als Vokabellisten im Bilingualunterricht.

Größter Wert wurde auf die Verwendung der korrekten britisch-englischen Fachbegriffe und Schreibweisen gelegt. Wo Begriffe aus dem amerikanischen Englisch gebräuchlich sind, werden aber auch diese aufgeführt mit dem Vermerk, dass es sich um einen amerikanisch-englischen Fachbegriff handelt.

Benutzungshinweise

Den englischen bzw. deutschen Begriffen sind zum Teil Hinweise beigefügt, die durch eckige Klammern als solche gekennzeichnet sind:

Beispiel	*Hinweis*	*Erklärung*
geerdet earthed [B. E.], grounded [A. E.]	[B. E.] [A. E.]	„earthed" in der Bedeutung von „geerdet" ist britisches Englisch. „grounded" in der Bedeutung von „geerdet" ist amerikanisches Englisch.
Unterdruck vacuum, underpressure [rare]	[selten] [rare]	„Unterdruck" wird überwiegend mit „vacuum" übersetzt. „underpressure" ist ungebräuchlich bzw. wird nur sehr selten benutzt.
gross calorific value Brennwert, oberer Heizwert [alt]	[alt] [old]	Die korrekte Übersetzung für „gross calorific value" ist Brennwert. Die Bezeichnung „oberer Heizwert" ist veraltet und sollte nicht mehr benutzt werden.
nitrogen [N; Z=7] Stickstoff	[**; Z=*]	nitrogen (Stickstoff) ist ein chemisches Element mit dem Elementsymbol N und der Ordnungszahl Z = 7.
adjacent [M] Ankathete	[M]	Dies ist ein Fachbegriff der Mathematik.
Doppelbindung [C] double bond	[C]	Dies ist ein Fachbegriff der Chemie.
→ **three wire cable**	→	siehe Details unter „three wire cable".
Zoll [Einh.] inch **acre [unit]** Morgen	[Einh.] [unit]	Die deutsche Einheit „Zoll" heißt im Englischen „inch". Die englische Einheit „acre" wird mit „Morgen" ins Deutsche übersetzt.
senkrecht [adj.] perpendicular, vertical **senkrecht [adv.]** perpendicular, perpendicularly, vertically	[adj.] [adv.]	„senkrecht" als Adjektiv wird mit „perpendicular" oder „vertical" übersetzt. Als Adverb wird „senkrecht" mit „perpendicular", „perpendicularly" oder „vertically" übersetzt.

Teil I
Englisch - Deutsch

1st Kepler law 1. Kepler'sches Gesetz
1st law of thermodynamics 1. Hauptsatz der Thermodynamik
1st Newtonian law 1. Newton'sches Gesetz
2nd Kepler law 2. Kepler'sches Gesetz
2nd law of thermodynamics 2. Hauptsatz der Thermodynamik
2nd Newtonian law 2. Newton'sches Gesetz
3rd Kepler law 3. Kepler'sches Gesetz
3rd law of thermodynamics 3. Hauptsatz der Thermodynamik
3rd Newtonian law 3. Newton'sches Gesetz
ablation Ablation
absolute temperature absolute Temperatur
absolute zero absoluter Nullpunkt
absorbance Absorptionsgrad
absorbed dose Energiedosis
absorption Absorption
AC Wechselstrom
AC current Wechselstrom
AC motor Wechselstrommotor
AC voltage Wechselspannung
to accelerate beschleunigen
acceleration Beschleunigung
acceleration amplitude Beschleunigungsamplitude
acceleration due to gravity Fallbeschleunigung
acceleration unit Beschleunigungseinheit
acceleration work Beschleunigungsarbeit
acceleration-elongation relation Beschleunigungs-Elongationsbeziehung
accelerator Beschleuniger
accelerator ring Kreisbeschleuniger, Ringbeschleuniger
accommodation Akkommodation
acetone [C] Aceton
acetylene [C] Acetylen
acoustic colour Klangfarbe
acoustics Akustik
acre [unit] Morgen
actinium [Ac, Z=89] Actinium
actio=reactio Actio=Reactio
activity Aktivität
activity law Aktivitätsgesetz
to add [M] addieren
addend [M] Summand
addition [M] Addition
additive mixture additive Mischung
to adhere haften
adhesion Haftung
adiabatic coefficient Adiabatenkoeffizient
adjacent [M] Ankathete
to adjust justieren
aerometer Aerometer
aeroplane [B. E.] Flugzeug
against the grain quer zur Faser

to age altern
aging Altern
air Luft
air bearing Luftkissen
air bearing stage Luftkissentisch
air circulation Luftzirkulation
air cushion Luftkissen
air pressure Luftdruck
air pump Luftpumpe
air resistance Luftwiderstand
airplane [A. E.] Flugzeug
ALCI (appliance leakage current interrupter) Fehlerstromschutzschalter, FI-Schutzschalter, FI-Sicherung
algebra [M] Algebra
to align ausrichten
allowed orbit erlaubte Bahn
alloy Legierung
along the grain parallel zur Faser
alpha decay Alphazerfall
alpha particle Alphateilchen
alpha radioactivity Alpharadioaktivität
alternating current Wechselstrom
alternating voltage Wechselspannung
altimeter Höhenmesser
aluminium [Al, Z=13] Aluminium
aluminized aluminiert
amber Bernstein
americium [Am, Z=95] Americium
ammeter Amperemeter
ammonia [C] Ammoniak
Amonton's law Amonton'sches Gesetz
amount of substance Stoffmenge
amp [unit] Ampere
ampere [unit] Ampere
amperemeter Amperemeter
amplitude Amplitude
analysis Analyse
analysis [M] Analysis
to anchor verankern
aneroid barometer Dosenbarometer
aneroid capsule (of an aneroid barometer) Vidi-Dose (eines Dosenbarometers)
aneroid cell (of an aneroid barometer) Vidi-Dose (eines Dosenbarometers)
angle [M] Winkel
angstrom [unit] Angström
angular frequency Kreisfrequenz
angular speed Winkelgeschwindigkeit
angular velocity Vektor der Winkelgeschwindigkeit
anion [C] Anion
anode Anode
anti-clockwise [adj.] [adv.] [B. E.] im Gegenuhrzeigersinn
antimony [Sb, Z=51] Antimon

antinode Wellenbauch
anvil (ear) Amboss (Ohr)
anvil (tool) Amboss (Werkzeug)
aperture Apertur, Blende
appliance leakage current interrupter Fehlerstromschutzschalter, FI-Schutzschalter, FI-Sicherung
application Anwendung
to apply voltage Spannung anlegen
to apply voltage to a resistor Spannung an einen Widerstand anlegen
approximate formula Näherungsformel
approximation Näherung
aquarium Aquarium
arc length [M] Bogenlänge
to arch sich wölben, wölben
arched gewölbt
arcus [M] arcus
are [unit] Are
area Fläche
argon [Ar, Z=18] Argon
armature Rotor
arrow Pfeil
arrow aperture Pfeilblende
arsenic [As, Z=33] Arsen
artificial radiation source künstliche Quelle radioaktiver Strahlung
asphalt Asphalt
to assemble zusammenbauen
assimilation Assimilation
astatine [At, Z=85] Astat
asteroid Asteroid
asteroid belt Asteroidengürtel
astronaut Astronaut
astronomical unit [unit] astronomische Einheit
astronomy Astronomie
atmosphere [unit] Atmosphäre
atmosphere (e. g. of Earth) Atmosphäre (z. B. der Erde)
atom Atom
atomic lattice Atomgitter
atomic mass unit atomare Masseneinheit
atomic nuclei Atomkerne
atomic nucleus Atomkern
atomic number [C] Ordnungszahl
atomic shell Atomhülle, Atomschale
atomic structure Atombau, Atomstruktur
atomic unit of charge [unit] atomare Ladungseinheit
to attenuate abklingen, sich abschwächen
to attract sich anziehen, anziehen
to attract each other sich gegenseitig anziehen
attraction Anziehung
attractive anziehend
aurora Polarlicht
aurora australis südliches Polarlicht, Südlicht
aurora borealis nördliches Polarlicht, Nordlicht
on average im Mittel
to average mitteln
average Mittelwert
average acceleration Durchschnittsbeschleunigung
average acceleration vector Vektor der Durchschnittsbeschleunigung
average mass number [C] mittlere Massenzahl
average speed Durchschnittsgeschwindigkeit
average velocity Vektor der Durchschnittsgeschwindigkeit
Avogadro's number Avogadro-Zahl
x-axis [M] x-Achse
y-axis [M] y-Achse
axis [M] Achse
axle Achse
background radiation natürliche Radioaktivität
balance Waage
balance of forces Kräftegleichgewicht
balance wheel Unruh (einer Uhr)
ball Kugel
ball bearing Kugellager
banana plug Bananenstecker
bar [unit] Bar
bar Stab, Stange
bar magnet Stabmagnet
barium [Ba, Z=56] Barium
barometer Barometer
barometric formula barometrische Höhenformel
barrel [unit] Barrel
base (of a power) [M] Basis (einer Potenz)
base (of a stand) Sockel (eines Stativs)
basic magnet Elementarmagnet
basis Grundlage
bass guitar Bassgitarre
bass string Basssaite
battery Batterie
beaker Becherglas
beat Schwebung
beat frequency Schwebungsfrequenz
becquerel [unit] Becquerel
beech wood Buchenholz
belt Riemen
bend Kurve
benzene [C] Benzol
berkelium [Bk, Z=97] Berkelium
beryllium [Be, Z=4] Beryllium
beta decay Betazerfall
beta factor Beta-Faktor
beta particle Betateilchen
beta radioactivity Betaradioaktivität
bicycle dynamo Fahrraddynamo

bicycle pump Fahrradpumpe
bimetallic thermometer Bimetallthermometer
binding forces [C] Bindungskräfte
binoculars Fernglas
birch wood Birkenholz
bismuth [Bi, Z=83] Bismut
black hole schwarzes Loch
black ice Glatteis
to blast sprengen
block Klotz
to blow sprengen
blow pipe Blasrohr
to blow up sprengen
blowgun Blasrohr
blueshift Blauverschiebung
to blur verschwimmen
blurred unscharf, verschwommen
bob Pendelkörper
body Körper, Objekt
body colours Körperfarben
bohrium [Bh, Z=107] Bohrium
boiling point Siedepunkt
bolt Bolzen (mit Gewinde), Metallschraube
Boltzmann constant Boltzmann-Konstante
bond [C] Bindung
bone Knochen
boron [B, Z=5] Bor
to bounce federn
bound electron gebundenes Elektron
boundary Grenzfläche
bow and arrow Pfeil und Bogen
boxcar Güterwagen
Boyle-Mariotte experiment Boyle-Mariotte-Experiment
Boyle-Mariotte's law Boyle-Mariotte'sches Gesetz
brake Bremse
to brake bremsen
brake cylinder Bremszylinder
brake disk Bremsscheibe
brake drum Bremstrommel
brake fluid Bremsflüssigkeit
brake hose Bremsschlauch
brake line Bremsleitung
brake pad Bremsbelag
brake pedal Bremspedal
brake shoe Bremsschuh
braking Bremsen
braking distance Bremsweg
braking time Bremszeit
branch Zweig
brass Messing
to break zerbrechen
bridge (guitar) Steg (Gitarre)
bright hell

brightness Helligkeit
brittle spröde
bromine [Br, Z=35] Brom
brush Bürste
brush (motor, generator) Bürste (Motor, Generator)
buffer Puffer
bullet Kugel (Geschoss)
bumper Stoßstange
buoyancy Auftrieb, Auftriebskraft
burner Brenner
to bust sprengen
butane [C] Butan
C-14 dating C-14-Methode
C-14 portion C-14-Anteil
cadmium [Cd, Z=48] Kadmium
caesium [Cs, Z=55] Cäsium
calcium [Ca, Z=20] Calcium
calculus [M] Differenzial- und Integralrechnung
to calibrate kalibrieren
calibrated kalibriert
californium [Cf, Z=98] Californium
caliper Messschieber, Schieblehre, Schublehre
calliper Messschieber, Schieblehre, Schublehre
calorie [unit] Kalorie
calorific value Brennwert, Heizwert
cam Nocke
cam belt Zahnriemen
cam belt pulley Zahnriemenscheibe
cam pulley Zahnriemenscheibe
cambelt Zahnriemen
camshaft Nockenwelle
to cancel each other out sich gegenseitig auslöschen
candela [unit] Candela
candle holder Kerzenhalter
capacitance Kapazität (eines Kondensators)
capacitor Kondensator
capacitor plate Kondensatorplatte
capacity (of a battery) Kapazität (einer Batterie)
capsule Kapsel
carat [unit] Karat
carbon Kohle
carbon [C, Z=6] Kohlenstoff
carbon dioxide [C] Kohlendioxid
carbon monoxide [C] Kohlenmonoxid
carbon steel Karbonstahl
carbon tetrachloride [C] Tetrachlorkohlenstoff
carbon-carbon bond [C] C-C-Bindung
carburetor [A. E.] Vergaser
carburettor [B. E.] Vergaser
cardiac fibrillation Herzflimmern
cargo Fracht

carpet cleaner [A. E.] Staubsauger
carrier medium Trägermedium
carrier of energy Energieträger
to cast a shadow einen Schatten werfen
cast iron Grauguss
cast steal Gusseisen
castor oil Rizinusöl
catapult seat Schleudersitz
caterpillar Raupe
cathode Kathode
cathode ray tube Kathodenstrahlrohr
cation [C] Kation
cause, transmission, effect Ursache, Vermittlung, Wirkung
cedar oil Zedernholzöl, Zedernöl
celerity Phasengeschwindigkeit
celestial object Himmelskörper
Celsius temperature Celsius-Temperatur
centimetre grid Zentimeterraster
central temperature (of a star) Zentraltemperatur (eines Sterns)
centre ray Zentralstrahl
centric zentrisch
centrifugal acceleration Zentrifugalbeschleunigung
centrifugal force Zentrifugalkraft
centrifuge Zentrifuge
centripetal acceleration Zentripetalbeschleunigung
centripetal force Zentripetalkraft
ceramics Keramik
cerium [Ce, Z=58] Cer
chain Kette
chamber pitch Kammerton
characteristic Kennlinie
characteristic curve Kennlinie
to charge aufladen, laden
charge Ladung
charge number Ladungszahl
charge-to-mass ratio for the electron Ladungs-Massenverhältnis des Elektrons
charging Aufladen, Aufladung
charging-up Aufladung
chart [M] Diagramm
chart Messdiagramm, Messtabelle
chemical bond [C] chemische Bindung
chemical symbol [C] Elementsymbol
chemistry Chemie
Chladni patterns Chladni'sche Klangfiguren
chlorine [Cl, Z=17] Chlor
chromium [Cr, Z=24] Chrom
ciliary muscle Ziliarmuskel
circle [M] Kreis
circuit diagram Schaltbild, Schaltplan
circuit symbol Schaltsymbol
circular current Kreisstrom
circular measure [M] Bogenmaß
circular motion Kreisbewegung
circular path Kreisbahn
circular wave Kreiswelle
circumference (especially of a circle) Umfang
classical relativity klassische Relativität
clay Ton
clay brick Ziegelstein
clock Uhr
clockwise [adj.] [adv.] im Uhrzeigersinn
closed circuit geschlossener Stromkreis
closed electric circuit geschlossener Stromkreis
closed system abgeschlossenes System
closest focusing distance minimale Sehweite, minimale Sichtdistanz
cloud chamber Nebelkammer
clutch Kupplung
clutch disk Kupplungsscheibe
coal-fired power station Kohlekraftwerk
cobalt [Co, Z=27] Kobalt
cochlea Gehörschnecke
cochlear nerve Hörnerv
cogwheel Zahnrad
coil Spule
to coil itself up sich aufwickeln
coil length Spulenlänge
coil spring Schraubenfeder
to coil up something etwas aufwickeln
collimation Kollimation
collimator Kollimator
collision Zusammenstoß
colour Farbe
colour addition Farbaddition
colour code Widerstandscode
colour filter Farbfilter
colour subtraction Farbsubtraktion
coloured light farbiges Licht
two colours form a third zwei Farben ergeben eine dritte
two colours produce a third zwei Farben ergeben eine dritte
two colours yield a third zwei Farben ergeben eine dritte
column of mercury Quecksilbersäule
to combine with [C] sich verbinden mit
combined [C] gebunden
combined with the lattice [C] gebunden an das Gitter
combustion Verbrennung
combustion chamber Brennkammer
comet Komet
communicating vessels kommunizierende Röhren
commutator Kommutator

compass Kompass, Zirkel
compass needle Kompassnadel
compasses Zirkel
complementary colours Komplementärfarben
complex number [M] komplexe Zahl
component Bauteil
compound Präparat
to compress zusammendrücken
compressed air Druckluft
compressibility Kompressibilität
compression (stroke 2 of a four-stroke engine) Verdichten (Takt 2 eines Viertaktmotors)
compression disturbance Verdichtungsstörung
compression spring Druckfeder
concave lens Konkavlinse
concave mirror Hohlspiegel
concave mirror equation Hohlspiegelgleichung
concept Konzept
to conclude schlussfolgern
concrete Beton
concurrency Gleichzeitigkeit
condensation Kondensation
condensation nucleus Kondensationskern
to condense kondensieren
to conduct (electrically) leiten (elektrisch)
to conduct (thermally) leiten (thermisch)
conducting wire Leiterdraht
conduction (electrical) Leitung (elektrisch)
conduction (thermal) Leitung (thermisch)
conductor (electrical) Leiter (elektrisch)
conductor (thermal) Leiter (thermisch)
conductor swing Leiterschaukel
cones (eye) Zapfen (Auge)
connected in parallel parallel geschaltet
connected in series in Reihe geschaltet, seriell geschaltet
constant function [M] konstante Funktion
constant of proportionality Proportionalitätskonstante
constantan Konstantan
constructive interference konstruktive Interferenz
to consume energy Energie konsumieren
to consume power Leistung konsumieren
contact area Berührungsfläche
contact force Kontaktkraft
convection Konvektion, Wärmeströmung
convection tube Konvektionsrohr
conventional current direction konventionelle Stromrichtung, technische Stromrichtung
convergent konvergent
converging lens Sammellinse

convex lens Konvexlinse
convex mirror Wölbspiegel
conveyor belt Förderband
cooker [B. E.] Herd
coordinate system [M] Koordinatensystem
copper [Cu, Z=29] Kupfer
Coriolis force Corioliskraft
cork Kork
cornea (eye) Hornhaut (Auge)
cosine [M] Cosinus
cosinus [M] Cosinus
cosmic radiation kosmische Strahlung
cotton cloth Baumwolltuch
cotton wool Watte
coulomb [unit] Coulomb
Coulomb force Coulombkraft
Coulomb force vector Vektor der Coulombkraft
Coulomb's law Coulomb-Gesetz
count rate Zählrate
counter field Gegenfeld
counterclockwise [adj.] [adv.] [A. E.] im Gegenuhrzeigersinn
counterforce Gegenkraft
counter-voltage Gegenspannung
coupled oscillators gekoppelte Oszillatoren
coupling Kopplung
course (e. g. of a voltage) Verlauf (z. B. einer Spannung)
covalent bond [C] kovalente Bindung
to crack zerbrechen
craftsman Handwerker
cramp Krampf
crane Kran
crank Kurbel
to crank kurbeln
crankshaft Kurbelwelle
crest Wellenberg
critical angle of total reflection kritischer Winkel der Totalreflexion, Winkel der Totalreflexion
crocodile clamp Krokodilklemme
cross sectional area Querschnittsfläche
crude oil Rohöl
crystal Kristall
crystal lattice Kristallgitter
cube [M] Würfel
cubic metre [unit] Kubikmeter
cubic metre per second [unit] Kubikmeter pro Sekunde
cuboid [M] Quader
cupola Kuppel
curie [unit] Curie
Curie temperature Curietemperatur
curium [Cm, Z=96] Curium
current Strom

current divider Stromteiler
current-voltage characteristic Strom-Spannungs-Kennlinie
current-voltage characteristic curve Strom-Spannungs-Kennlinie
curve Kurve
curved gewölbt
to cut a thread ein Gewinde schneiden
cyan cyan
cylinder [M] Zylinder
cylinder (of an engine) Zylinder (eines Motors)
cylinder capacity Hubraum
cylinder head Zylinderkopf
dam Staudamm
damped oscillation gedämpfte Schwingung
damping Dämpfung
damping coefficient Dämpfungsfaktor
danger of death Todesgefahr
dark dunkel
data point Datenpunkt
daughter nucleus Tochterkern
day [unit] Tag
DC Gleichstrom
DC current Gleichstrom
DC motor Gleichstrommotor
DC voltage Gleichspannung
de Broglie relation de Broglie-Beziehung
decay Zerfall
to decay zerfallen
to decay (oscillation) abklingen, sich abschwächen
decay chain Zerfallsreihe
decay constant Zerfallskonstante
decay mode Zerfallsmodus
deceleration Bremsverzögerung, Verzögerung
decimal place [M] Dezimalstelle
decimal point [M] Dezimalpunkt
declination Deklination
to decompose a vector [M] einen Vektor zerlegen
decomposition radioaktiver Zerfall, Zerfall
decrease Abfall, Abnahme
to deduce ableiten (logisch), herleiten
deduction Herleitung
deflection Auslenkung, Elongation
to deform deformieren, verformen
deformation Deformation, Verformung
degree [M] Grad
degree Celsius [unit] Grad Celsius
degree Fahrenheit [unit] Grad Fahrenheit
denominator [M] Nenner
density Dichte
dependence of A on B Abhängigkeit von A von B
to depict bildlich darstellen, darstellen

to deposit resublimieren
deposition Resublimation
depth Tiefe
to derail entgleisen
derivation [M] Ableitung
to derive ableiten
to derive [M] ableiten, differenzieren
derived SI-Unit abgeleitete SI-Einheit
desirable friction erwünschte Reibung
destructive interference destruktive Interferenz
deterioration Abnutzung
diagram [M] Diagramm
diagram Messdiagramm
diamagnet Diamagnet
diamagnetic diamagnetisch
diamagnetic substance diamagnetische Substanz
diamagnetism Diamagnetismus
diamond Diamant
dielectric Dielektrikum
dielectrics Dielektrika
diesel Diesel
diethyl ether [C] Diethylether
difference [M] Differenz
differential calculus [M] Differenzialrechnung
diffraction Beugung
diffraction angle Beugungswinkel
diffraction grating Beugungsgitter, Gitter
diffuse reflection diffuse Reflexion
diffuser Diffusor
diffusion (of a wave) Ausbreitung (einer Welle)
digit [M] Ziffer
dimension Dimension
dimmer Dimmer
diode Diode
dioptre Dioptrie
dipole Dipol
direct current Gleichstrom
direct current voltage Gleichspannung
direct voltage Gleichspannung
direction Richtung
direction of motion Bewegungsrichtung
direction of rotation Drehsinn, Rotationsrichtung
disintegration radioaktiver Zerfall, Zerfall
disk brake Scheibenbremse
dispersion Dispersion, Streuung
dispersion (chromatic dispersion of light) Dispersion (chromatische Dispersion von Licht)
to displace verdrängen
to displace [M] verschieben
to displace water Wasser verdrängen
displaced verdrängt
displaced [M] verschoben
displaced liquid verdrängte Flüssigkeit

displacement Abstandsvektor
displacement [M] Verschiebungsvektor
distilled water destilliertes Wasser
disturbance Störung
divergent divergent
diverging lens Zerstreuungslinse
to divide [M] dividieren
a divided by b [M] a geteilt durch b
division [M] Division
to do work Arbeit verrichten
to double verdoppeln
double bond [C] Doppelbindung
double glazing Doppelverglasung
double refraction Doppelbrechung
double slit Doppelspalt
double-walled doppelwandig
drag Strömungswiderstand
drag coefficient Widerstandsbeiwert
drill Bohrer, Bohrmaschine
drive belt Antriebsriemen
drive pulley Antriebsriemenscheibe, Antriebsscheibe
driven axle angetriebene Achse
drum brake Trommelbremse
dry ice Trockeneis
dubnium [Db, Z=105] Dubnium
dwarf planet Zwergplanet
dynamic friction Gleitreibung
dynamic friction coefficient Gleitreibungszahl
dynamic friction force Gleitreibungskraft
dynamic pressure Staudruck
dynamic viscosity dynamische Viskosität
dynamics Dynamik
dynamo Dynamo, Lichtmaschine (Auto)
dynamo-electric principle dynamoelektrisches Prinzip
dysprosium [Dy, Z=66] Dysprosium
ear Ohr
ear canal Gehörgang
ear muffs Ohrenschützer
ear ossicles Gehörknöchelchen
ear plugs Ohrenstopfen, Ohrenstöpsel
eardrum Trommelfell (Ohr)
Earth Erde
earth (electric wire) [B. E.] Erde (elektrisches Kabel)
to earth [B. E.] erden
earth wire [B. E.] Erdkabel
earthed [B. E.] geerdet
earthing [B. E.] Erden, Erdung
earthquake Erdbeben
Earth's core Erdkern
Earth's crust Erdkruste
Earth's magnetic field Erdmagnetfeld
Earth's mantle Erdmantel
eccentric exzentrisch
eccentricity Exzentrizität
eddy current Wirbelstrom
eddy current brake Wirbelstrombremse
edge length [M] Kantenlänge
efficiency Effizienz, Wirkungsgrad
eigenfrequency Eigenfrequenz
einsteinium [Es, Z=99] Einsteinium
ejection seat Schleudersitz
elastic deformation elastische Deformation
elasticity Elastizität
electric charge elektrische Ladung
electric circuit elektrischer Stromkreis
electric constant Dielektrizitätskonstante des Vakuums [alt], elektrische Feldkonstante, Permittivität des Vakuums [selten]
electric current elektrische Stromstärke, elektrischer Strom
electric field elektrisches Feld
electric field constant [rare] Dielektrizitätskonstante des Vakuums [alt], elektrische Feldkonstante, Permittivität des Vakuums [selten]
electric field line elektrische Feldlinie
electric field strength elektrische Feldstärke
electric force elektrische Kraft
electric motor Elektromotor
electric power elektrische Leistung
electric radiator elektrischer Heizkörper
electric resistance elektrischer Widerstand (physikalische Größe)
electric resistor elektrischer Widerstand (Bauteil)
electric shock elektrischer Schlag, Stromschlag
electrical component elektrisches Bauteil
electrical resistance elektrischer Widerstand (physikalische Größe)
electrical resistor elektrischer Widerstand (Bauteil)
electrical work Arbeit des elektrischen Stroms
electricity Elektrizität
electrode Elektrode
electrodynamics Elektrodynamik
electromagnetic force elektromagnetische Kraft
electromagnetic induction Induktion
electromagnetic spectrum elektromagnetisches Spektrum
electromagnetism Elektromagnetismus
electromotive force Biot-Savart-Kraft
electron Elektron
electron avalanche Elektronenlawine
electron density Elektronendichte
electron excitation Elektronenanregung
electron number Elektronenzahl
electron volt [unit] Elektronenvolt
electronic signal elektronisches Signal

electroscope Elektroskop
electrostatic induction Influenz
electrostatics Elektrostatik
element [C] Element
element name [C] Elementname
elementary charge Elementarladung
elevator [A. E.] Aufzug
elevator cab [A. E.] Aufzugkabine
elevator cabin [A. E.] Aufzugkabine
ellipse Ellipse
elliptic elliptisch
elongation Auslenkung, Elongation
emission Emission
emissivity Emissivität
to emit emittieren
to enclose an angle with [M] einen Winkel einschließen mit
to encompass an angle with [M] einen Winkel einschließen mit
encrypted verschlüsselt
end stop (e. g. of a spring scale) Anschlag (z. B. einer Federwaage)
energy Energie
energy carrier Energieträger
energy conservation law Energieerhaltungssatz
energy density Energiedichte
energy flow Energiefluss
energy gain Energieausbeute, Energiegewinn
energy level Energieniveau
energy loss Energieverlust
energy transfer Energieübertragung
energy transformation Energieumwandlung
energy transformation chain Energieumwandlungskette
energy yield Energieausbeute, Energiegewinn
energy-rich energiereich
energy-saving bulb Sparlampe
engine Motor
enlarged image vergrößertes Bild
epicentre Epizentrum
to equal [M] gleich sein
equation [M] Gleichung
equation Reaktionsgleichung
equation of motion Bewegungsgleichung
equatorial radius Äquatorradius
equatorial surface gravity Fallbeschleunigung an der Oberfläche am Äquator
equidistant äquidistant
equilibrium Gleichgewicht
equilibrium of forces Kräftegleichgewicht
equilibrium position Gleichgewichtslage
equipment Ausrüstung
equivalent dose Äquivalentdosis, Äquivalenzdosis
erbium [Er, Z=68] Erbium
Erlenmeyer flask Erlenmeyerkolben
to estimate abschätzen
estimation Abschätzung
ethane [C] Ethan
ethanol [C] Ethanol
European integrated network Europäisches Verbundnetz
europium [Eu, Z=63] Europium
Eustachian tube (ear) eustachische Röhre (Ohr), Ohrtrompete (Ohr)
to evaluate auswerten
evaluation Auswertung
excavator Bagger
excavator shovel Baggerschaufel
excitance (power per surface area) Ausstrahlung (Leistung pro Oberfläche)
excitation (atom, electron, nucleus) Anregung (Atom, Elektron, Atomkern)
to excite (atom, electron, nucleus) anregen (Atom, Elektron, Atomkern)
to excite (oscillation) anregen (Schwingung)
to excite (wave) anregen (Welle)
excited (atom, electron, nucleus) angeregt (Atom, Elektron, Atomkern)
excited state angeregter Zustand
exciter (of a wave) Erreger (einer Welle)
exciter (of an oscillation) Erreger (einer Schwingung)
to exert a force eine Kraft ausüben
to exert pressure Druck ausüben
exhaust (stroke 4 of a four-stroke engine) Ausstoßen (Takt 4 eines Viertaktmotors)
expansion Expansion
experiment Experiment
experimental setup Versuchsaufbau
experimental set-up Versuchsaufbau
explosion Explosion
exponent [M] Exponent
exponential function [M] Exponentialfunktion
exponentiation [M] Potenzieren
extension Ausdehnung
extension spring Zugfeder
exterior Äußeres
external combustion externe Verbrennung
to extract a root [M] radizieren, eine Wurzel ziehen
eye Auge
eye lens Augenlinse
eyeball Augapfel
eyepiece Okular
factor [M] Faktor
to factorize [M] ausklammern, faktorisieren
farad [unit] Farad
Faraday cage Faraday-Käfig
fat droplet Fetttröpfchen
fauna Fauna

fermium [Fm, Z=100] Fermium
ferromagnet Ferromagnet
ferromagnetic ferromagnetisch
ferromagnetic substance ferromagnetische Substanz
ferromagnetism Ferromagnetismus
fibreglass Glasfaser
field of view Sehwinkel
field strength Feldstärke
filament (of a light bulb) Glühfaden (einer Glühbirne)
final point of a vector [M] Endpunkt eines Vektors
fire hose Feuerwehrschlauch
first Kepler law erstes Kepler'sches Gesetz
first law of thermodynamics erster Hauptsatz der Thermodynamik
first Newtonian law erstes Newton'sches Gesetz
fissile spaltbar
fit Anpassungskurve, Fit
to fit fitten
fitting Fitten
fixed end festes Ende
flat battery leere Batterie
flat tyre platter Reifen
flight time Flugdauer
flint Feuerstein
flint glass Flintglas
to float schweben (in der Luft), schweben (unter Wasser), schwimmen (Objekte oder Lebewesen, die auf dem Wasser treiben), treiben (Objekte oder Lebewesen auf dem Wasser)
flora Flora
flow Fluss
fluid Fluid
fluid resistance Strömungswiderstand
fluorescent lamp Leuchtstoffröhre
fluorine [F, Z=9] Fluor
flywheel Schwungrad
focal length Brennweite
focal plane Brennebene
focal point Brennpunkt
focal ray Brennstrahl
focus Brennpunkt
foot [unit] Fuß
force Kraft
forced oscillation erzwungene Schwingung
fossil fossil
Fourier analysis [M] Fourier-Analyse
four-stroke engine Viertaktmotor
fraction [M] Bruch
fraction bar [M] Bruchstrich
francium [Fr, Z=87] Francium
Fraunhofer line Fraunhoferlinie
free electron freies Elektron

free end freies Ende
free fall freier Fall
to freeze erstarren
freezing Gefrieren
freezing point Gefrierpunkt
freon [C] Freon
frequency Frequenz
frequency component Frequenzanteil
fresh water Süßwasser
fret (guitar) Bund (Gitarre), Bundstab (Gitarre)
friction Reibung
friction coefficient Reibungszahl
friction work Reibungsarbeit
frictionless reibungsfrei
fuel Treibstoff
fuel gauge Tankuhr
full circle [M] Vollkreis
full solar eclipse totale Sonnenfinsternis
function principle Funktionsprinzip
function value [M] Funktionswert
functionality Funktionsweise
fundamental Grundschwingung, Grundton
fundamental constant fundamentale Konstante
fundamental oscillation Grundschwingung
fundamental tone Grundton
to fur up verkalken
fuse Sicherung
a fuse blows eine Sicherung fliegt heraus
a fuse is blown eine Sicherung ist herausgeflogen
gadget Gerät
gadolinium [Gd, Z=64] Gadolinium
Galilean telescope Galileisches Fernrohr
Galilean transformation Galilei-Transformation
Galileo's telescope Galileisches Fernrohr
gallium [Ga, Z=31] Gallium
gallon [unit] Gallone
gamma decay Gammazerfall
gamma particle Gammateilchen
gamma radioactivity Gammaradioaktivität
gas Gas
gas [A. E.] Benzin
gas constant allgemeine Gaskonstante
gas cylinder Gasflasche
gas oil Zweitaktöl
gas particle Gasteilchen
gaseous gasförmig
gasoline [A. E.] Benzin
gauge Messschieber, Schieblehre, Schublehre
Gay-Lussac's law Gay-Lussac'sches Gesetz
Geiger-Müller tube Geiger-Müller-Zähler
general theory of relativity allgemeine Relativitätstheorie
generalisation Verallgemeinerung

generator Generator
generator current Generatorstrom
generator voltage Generatorspannung
genuine part Originalteil
geomagnetism Erdmagnetismus, Geomagnetismus
geometric object [M] geometrisches Objekt
geometry [M] Geometrie
geostationary geostationär
geothermal power station geothermisches Kraftwerk
germanium [Ge, Z=32] Germanium
germanium diode Germaniumdiode
GFCI (ground fault circuit interrupter) Fehlerstromschutzschalter, FI-Schutzschalter, FI-Sicherung
glass Glas
glass tube Glasrohr
glasses Brille
glider Segelflugzeug
globe Globus
to glow glühen
glow cathode Glühkathode
glow lamp Glimmlampe
glycerin Glycerin, Glycerol, Glyzerin
glycerine Glycerin, Glycerol, Glyzerin
glycerol Glycerin, Glycerol, Glyzerin
to go around a bend um eine Kurve fahren
to go round a curve um eine Kurve fahren
gold [Au, Z=79] Gold
gradient [M] Steigung
gradient (of an inclined plane) Steigung (einer schiefen Ebene)
grain [unit] Gran
granite Granit
graph [M] Diagramm, Graph, Kurve
graph Messdiagramm
graphite Graphit
grating Beugungsgitter, Gitter
gravel Kies, Schotter
gravitation Gravitation, Schwerkraft
gravitational constant Gravitationskonstante
gravitational field strength Fallbeschleunigung, Gravitationsfeldstärke
gravitational force Gravitationskraft, Schwerkraft
gravitational force vector Vektor der Gravitationskraft
gravity Gravitation (bezogen auf die Erde), Schwerkraft (bezogen auf die Erde)
gray [unit] Gray
grease Fett
greasy fettig, glitschig
great ocean conveyor belt Globales Förderband
grid [B. E.] Stromnetz
to grind schleifen
grinding stone Schleifstein
grindstone Schleifstein
grit Splitt
gross calorific value Brennwert, oberer Heizwert [alt]
gross heat of combustion Brennwert, oberer Heizwert [alt]
ground (electric wire) [A. E.] Erde (elektrisches Kabel)
to ground [A. E.] erden
ground [A. E.] Erdung
ground fault circuit interrupter Fehlerstromschutzschalter, FI-Schutzschalter, FI-Sicherung
ground level Grundzustand
ground state Grundzustand
ground wire [A. E.] Erde (elektrisches Kabel)
grounded [A. E.] geerdet
grounding [A. E.] Erden, Erdung
guitar Gitarre
guitar string Gitarrensaite
Gulf Stream Golfstrom
hafnium [Hf, Z=72] Hafnium
half life Halbwertszeit
half ring Halbring
Hall effect Hall-Effekt
Hall probe Hall-Sonde
halogen lamp Halogenlampe
to halve halbieren
hammer (ear) Hammer (Ohr)
hammer (tool) Hammer (Werkzeug)
hammer handle Hammerstiel
hammer head Hammerkopf
handbrake Handbremse
handlebars Lenkstange (Fahrrad)
hard rubber Hartgummi
harmonic Oberschwingung
harmonic oscillation harmonische Schwingung
harmonic wave harmonische Welle
hassium [Hs, Z=108] Hassium
to have an X-ray sich röntgen lassen
head light (car) Scheinwerfer (Auto)
hearing damage Gehörschädigung
hearing threshold Hörschwelle
heartbeat Herzschlag
heat Wärme
heat bridge Wärmebrücke
heat capacity Wärmekapazität
heat conduction Wärmeleitung
heat conduction coefficient Wärmeleitfähigkeitskoeffizient
heat engine Wärmekraftmaschine
heat of combustion Brennwert, Heizwert
heat propagation Wärmeausbreitung
heat pump Wärmepumpe

heat transfer Wärmedurchgang
heat transfer coefficient
Wärmedurchgangskoeffizient
heating value Brennwert, Heizwert
heating wire Heizdraht
heavy damping starke Dämpfung
heavy water schweres Wasser
hectare [unit] Hektar
height Höhe
Heisenberg uncertainty principle
Heisenberg'sche Unschärferelation
helical schraubenförmig
helical path Schraubenbahn
helium [He, Z=2] Helium
helix Schraubenlinie
helm Steuerruder
henry [unit] Henry
heptane [C] Heptan
hertz [unit] Hertz
hexane [C] Hexan
high tide Flut
high voltage Hochspannung
higher heating value Brennwert, oberer Heizwert [alt]
hoist Flaschenzug
holder (e. g. for a slit aperture) Halter (z. B. für eine Spaltblende)
holmium [Ho, Z=67] Holmium
homogeneous homogen
homogenous homogen
Hooke's law Hooke'sches Gesetz
Hoover [B. E.] Staubsauger
horizontal throw waagrechter Wurf
horsepower [unit] Pferdestärke
horseshoe magnet Hufeisenmagnet
hose Schlauch (groß)
hose pressure Schlauchdruck
hotplate Herdplatte
hour [unit] Stunde
housing Gehäuse
hovercraft Luftkissenschiff
humidity Luftfeuchtigkeit
hydraulic hydraulisch
hydraulic lift Hebebühne
hydraulic press hydraulische Presse
hydraulics Hydraulik
hydroelectric power station
Wasserkraftwerk
hydrogen [H, Z=1] Wasserstoff
hydrogen lamp Wasserstofflampe
hydrostatic paradox hydrostatisches Paradoxon
hydrostatic pressure Schweredruck
hydrostatics Hydrostatik
hyperbola [M] Hyperbel
hyperopic weitsichtig

hypocentre Hypozentrum
hypotenuse [M] Hypotenuse
hypotheses Hypothesen
hypothesis Hypothese
ice skate Schlittschuh
iceberg Eisberg
ideal gas equation allgemeine Gasgleichung
ignition plug Zündkerze
illuminance Beleuchtungsstärke
illumination Beleuchtung
image Bild
image distance Bildweite
image formation Abbildung
image formation equation
Abbildungsgleichung
image orientation Bildorientierung
image size Bildgröße
to immerse eintauchen
immersion heater Tauchsieder
impulse Kraftstoß
in a clockwise direction im Uhrzeigersinn
in a counterclockwise direction [A. E.]
im Gegenuhrzeigersinn
in an anti-clockwise direction [B. E.]
im Gegenuhrzeigersinn
in parallel in Parallel
in phase in Phase
in series in Reihe
incandescent bulb Glühbirne
incandescent light bulb Glühbirne
inch [unit] Zoll
incidence Einfall (von Licht)
incidence angle Einfallswinkel
incident ray einfallender Strahl
inclination Inklination
inclination (of an inclined plane) Neigung (einer schiefen Ebene), Neigungswinkel (einer schiefen Ebene)
inclination angle (of an inclined plane)
Neigungswinkel (einer schiefen Ebene)
inclined plane schiefe Ebene
incompressibility Inkompressibilität
increase Anstieg
index [M] Index, Wurzelexponent
indices [M] Indizes
indium [In, Z=49] Indium
induced voltage induzierte Spannung
inductance Induktivität
induction Induktion
induction cooker [B. E.] Induktionsherd
induction stove Induktionsherd
induction voltage Induktionsspannung
inertia Trägheit
inertial system Inertialsystem
to infer schlussfolgern
to inflate aufblasen

infrared light infrarotes Licht
infrasonic Infraschall-
infrasound Infraschall
initial activity anfängliche Aktivität
initial point of a vector [M] Anfangspunkt eines Vektors
initial speed Anfangsgeschwindigkeit
ink cartridge Druckerpatrone, Tintenpatrone (Drucker)
inner ear Innenohr
instantaneous acceleration Momentanbeschleunigung
instantaneous speed Momentangeschwindigkeit
instantaneous velocity Vektor der Momentangeschwindigkeit
to insulate (electrically) isolieren (elektrisch)
to insulate (thermally) isolieren (thermisch)
insulation (electrical) Isolation (elektrisch)
insulation (thermal) Isolation (thermisch)
insulator (electrical) Isolator (elektrisch)
insulator (thermal) Isolator (thermisch)
intake (stroke 1 of a four-stroke engine) Ansaugen (Takt 1 eines Viertaktmotors)
integral [M] Integral
integral calculus [M] Integralrechnung
to integrate [M] integrieren
integration [M] Integration
intensity Intensität
to interfere interferieren
interference Interferenz
interior Inneres
to interlock sich verzahnen, verzahnen
intermediate image Zwischenbild
internal combustion interne Verbrennung
internal energy innere Energie
internal resistance Innenwiderstand
internal resistance of the human body Innenwiderstand des menschlichen Körpers
international colour code internationaler Farbcode, internationaler Widerstandscode
intersection [M] Schnittfläche, Schnittpunkt
interval [M] Intervall
invar Invar
inverse Galilean transformation inverse Galilei-Transformation
inversely proportional umgekehrt proportional
inversely proportional to umgekehrt proportional zu
inverted image umgekehrtes Bild
to investigate untersuchen
inwards [adj.] [adv.] nach innen
iodine [I, Z=53] Iod
ion Ion
ion propulsion Ionenantrieb

ionic bond [C] Ionenbindung
ionisation Ionisation
iridium [Ir, Z=77] Iridium
iris Iris
iron Bügeleisen
iron [Fe, Z=26] Eisen
iron filings Eisenfeilspäne, Eisenspäne
iron girder Eisenträger
irradiance (power per area) Einstrahlung (Leistung pro Fläche)
irregular reflection unregelmäßige Reflexion
isobaric isobar
isochoric isochor
isothermal isotherm
isotope Isotop
jiggling motion Zitterbewegung
joule [unit] Joule
Julian year [unit] Julianisches Jahr
Jupiter Jupiter
KCL (Kirchhoff's current law) Kirchhoffsche Knotenregel, Knotenregel
KE (kinetic energy) kinetische Energie
kelvin [unit] Kelvin
Kepler law Kepler'sches Gesetz
Kepler telescope Kepler'sches Fernrohr
kerosene Kerosin
kettle Wasserkessel
kilogram [unit] Kilogramm
kilogram per cubic metre [unit] Kilogramm pro Kubikmeter
kilometre per hour [unit] Kilometer pro Stunde
kilowatt-hour [unit] Kilowattstunde
kinematics Kinematik
kinetic energy kinetische Energie
kinetic energy of a gas particle kinetische Energie eines Gasteilchens
Kirchhoff's circuit laws Kirchhoffsche Gesetze
Kirchhoff's current law Kirchhoffsche Knotenregel, Knotenregel
Kirchhoff's voltage law Kirchhoffsche Maschenregel, Maschenregel
kite Drachen
knitting needle Stricknadel
knot [unit] Knoten
krypton [Kr, Z=36] Krypton
Kuiper belt Kuiper-Gürtel
Kundt's tube Kundt'sches Rohr
KVL (Kirchhoff's voltage law) Kirchhoffsche Maschenregel, Maschenregel
lacquer Lack
Lagrangian point Lagrangepunkt
laminar flow laminare Strömung
lamp Lampe
lamp mounting Lampenaufhängung

lanthanum [La, Z=57] Lanthan
latent heat latente Wärme
latitude Breitengrad
lattice Gitter (eines Festkörpers), Kristallgitter
to launch starten
law of Archimedes Gesetz des Archimedes
law of gravitation Gravitationsgesetz
law of nature Naturgesetz
law of reflection Reflexionsgesetz
law of refraction Brechungsgesetz
lawrencium [Lr, Z=103] Lawrencium
lead [Pb, Z=82] Blei
leakage current Leckstrom
leather Leder
LED (light emitting diode) LED (lichtemittierende Diode), Leuchtdiode
legend (of a diagram) Legende (eines Diagramms)
length Länge
length contraction Längenkontraktion
length of a coil Spulenlänge
length of a vector [M] Länge eines Vektors
length of arc [M] Bogenlänge
lens Linse
lens combination Linsenkombination
lens distance Linsenabstand
lens equation Linsengleichung
lens system Linsensystem
Lenz's rule Lenz'sche Regel
level Wasserwaage
level plane waagrechte Ebene
levelling screw Stellschraube (zur Höheneinstellung)
to levitate schweben, schweben lassen
lid Deckel
lift [B. E.] Aufzug
lift car [B. E.] Aufzugkabine
lifting work Hubarbeit
light barrier Lichtschranke
light bulb Glühbirne
a light bulb shines eine Glühbirne leuchtet
light damping schwache Dämpfung
light emitting diode Leuchtdiode
light fixture Lampenaufhängung
light propagation Lichtausbreitung
light ray Lichtstrahl
light rays coincide Lichtstrahlen treffen aufeinander, Lichtstrahlen treffen sich
light source Lichtquelle
lightly damped oscillation schwach gedämpfte Schwingung
lightning Blitz
lightning conductor Blitzableiter
light-year [unit] Lichtjahr
like poles gleichnamige Pole
lime Kalk
limestone Kalkstein
line spectra Linienspektren
line spectrum Linienspektrum
linear accelerator Linearbeschleuniger
linear density (of a guitar string) Längendichte (einer Gitarrensaite)
linear function [M] lineare Funktion
linear motion lineare Bewegung
linear thermal expansion Längenausdehnung
linear thermal expansion coefficient Längenausdehnungskoeffizient
linseed oil Leinsamenöl
liquid flüssig, Flüssigkeit
liquid thermometer Flüssigkeitsthermometer
liquids Flüssigkeiten
lithium [Li, Z=3] Lithium
litre [unit] Liter
to load belasten
locomotion Fortbewegung
locomotive Lokomotive
logarithm [M] Logarithmus
logarithm of a to base b [M] Logarithmus von a zur Basis b
logical [M] logisch
longitude Längengrad
longitudinal coupling longitudinale Kopplung
longitudinal wave longitudinale Welle
long-range missile Langstreckenrakete
long-sighted weitsichtig
long-sightedness Weitsichtigkeit
loop Schleife
loop (of a coil) Windung (einer Spule)
Lorentz factor Lorentz-Faktor
Lorentz force Lorentzkraft
loudness Lautstärkepegel
loudspeaker Lautsprecher
low tide Ebbe
lower heating value Heizwert, unterer Heizwert [alt]
lubricant Schmiermittel
to lubricate schmieren
luminescent layer (of a fluorescent lamp) Leuchtschicht (einer Leuchtstoffröhre)
luminous intensity Luminosität
luminous object Selbstleuchter
lung Lunge
lutetium [Lu, Z=71] Lutetium
macrocosm Makrokosmos
macroscopic makroskopisch
Magdeburg sphere Magdeburger Kugel
magenta magenta
magma Magma
magnesium [Mg, Z=12] Magnesium
magnetic core Eisenkern
magnetic field Magnetfeld

magnetic field axis Achse des magnetischen Feldes
magnetic field line Magnetfeldlinie, magnetische Feldlinie
magnetic field strength Magnetfeldstärke, magnetische Feldstärke
magnetic flux magnetischer Fluss
magnetic flux density magnetische Flussdichte
magnetic force magnetische Kraft
magnetic induction magnetische Influenz
magnetic response magnetische Reaktion
magnetisation Magnetisierung
to magnetise magnetisieren
magnetised magnetisiert
magnetism Magnetismus
magnetite Magnetit
magnification Vergrößerung
magnifying glass Lupe
magnitude of a vector [M] Betrag eines Vektors, Länge eines Vektors
main decay mode Hauptzerfallsmodus
main maxima Hauptmaxima
main maximum Hauptmaximum
main switch Hauptschalter
mains Stromnetz
major scale Dur-Tonleiter
makeup Aufbau
manganese [Mn, Z=25] Mangan
manned mission bemannte Mission
to manoeuver manövrieren
manometer Manometer
to manufacture anfertigen
manufacture Anfertigung
to map [M] abbilden
mapping [M] Abbildung
marble Marmor
Mars Mars
mass Masse
mass number Massenzahl
mass spectrometer Massenspektrometer
mass-energy equivalence Masse-Energie-Äquivalenz
mast Mast
material control Materialprüfung
material property Materialeigenschaft
materials Materialien
mathematics Mathematik
matt matt
maxima Maxima
maximum Maximum
maximum efficiency maximaler Wirkungsgrad
maximum height maximale Höhe
mean radius mittlerer Radius
measurand Messgröße
to measure messen
measured data Messdaten, Messwerte
measurement Messung
measurement device Messgerät
measuring device Messgerät
measuring instrument Messgerät
measuring tape Maßband
mechanics Mechanik
meitnerium [Mt, Z=109] Meitnerium
to melt schmelzen
melting Schmelzen
melting point Schmelzpunkt
mendelevium [Md, Z=101] Mendelevium
Mercury Merkur
mercury [Hg, Z=80] Quecksilber
mercury thermometer Quecksilberthermometer
metallic bond [C] Metallbindung
metallic conductor metallischer Leiter
metallized verspiegelt
methane [C] Methan
methanol [C] Methanol
methods Methoden
metre [unit] Meter
metre per second [unit] Meter pro Sekunde
metre per second squared [unit] Meter pro Quadratsekunde
metric tonne [unit] Tonne
mica Glimmer
microcosm Mikrokosmos
microgravity Schwerelosigkeit
microphone Mikrophon
microscope Mikroskop
microscopic mikroskopisch
middle ear Mittelohr
middle ear cavity Paukenhöhle (Ohr)
mile [unit] Meile
mileometer Kilometerzähler
millimetre of mercury [unit] Millimeter Quecksilbersäule
milometer [rare] Kilometerzähler
minima Minima
minimisation Minimierung
to minimise minimieren
minimum Minimum
minuend [M] Minuend
minus [M] minus
a minus b [M] a minus b
a minus b equals c [M] a minus b ist gleich c
minus pole Minuspol
minute [unit] Minute
mirror Spiegel
mirrored verspiegelt
missile Rakete (Waffe)
to mix coloured light farbiges Licht mischen
model Modell
model concept Modellvorstellung
modelling clay Knete, Knetmasse

modulated moduliert
modulus of elasticity Elastizitätsmodul
moisture Feuchtigkeit
mole [unit] Mol
molten (e. g. iron) geschmolzen (z. B. Eisen)
molybdenum [Mo, Z=42] Molybdän
moment [rare] Drehmoment
momentum Impuls
moon Mond
motion Bewegung
motor Motor
motorcycle Motorrad
multiplication [M] Multiplikation
to multiply [M] multiplizieren
multi-slit aperture Mehrfachspaltblende
music Musik
musical scale Tonleiter
myopic kurzsichtig
natural background radiation natürliche Radioaktivität
natural frequency Eigenfrequenz
natural gas [C] Erdgas
natural number [M] natürliche Zahl
natural oscillation Eigenschwingung
natural science Naturwissenschaft
nature Natur
nautical mile [unit] nautische Meile
needle bearing Nadellager
negative pole negativer Pol
negative sign [M] negatives Vorzeichen
negatively charged negativ geladen
to neglect vernachlässigen
negligible vernachlässigbar
neodymium [Nd, Z=60] Neodym
neon [Ne, Z=10] Neon
neon lamp Neonlampe
Neptune Neptun
neptunium [Np, Z=93] Neptunium
net Netz
net calorific value Heizwert, unterer Heizwert [alt]
net heat of combustion Heizwert, unterer Heizwert [alt]
neutral neutral
neutral wire Nullleiter (das Kabel)
neutron Neutron
neutron number Neutronenzahl
neutron star Neutronenstern
newton [unit] Newton
newton metre [unit] Newtonmeter
Newtonian laws Newton'sche Gesetze
Newton's law of gravitation Newton'sches Gravitationsgesetz
nickel [Ni, Z=28] Nickel
niobium [Nb, Z=41] Niob

nitrogen [N, Z=7] Stickstoff
nobelium [No, Z=102] Nobelium
nodal line Knotenlinie
node Knoten, Wellenknoten
nomenclature Nomenklatur
nominal value Sollwert
nonane [C] Nonan
non-luminous object Fremdleuchter
non-SI-Unit Nicht-SI-Einheit
norm [M] Betrag
norm of a vector [M] Betrag eines Vektors, Länge eines Vektors
normal Normale
normal force Normalkraft
normal force vector Vektor der Normalkraft
North Pole Nordpol (der Erde)
north pole Nordpol (eines Magneten)
northern lights nördliches Polarlicht, Nordlicht
note (C, D, E ...) Note (c, d, e ...)
nozzle Düse
nuclear fission Kernspaltung
nuclear force Kernkraft
nuclear fusion Kernfusion
nuclear plant Kernkraftwerk
nuclear power plant Atomkraftwerk, Kernkraftwerk
nuclear power station Kernkraftwerk
nuclear radiation radioaktive Strahlung
nuclear warhead Atomsprengkopf
nuclei Atomkerne, Kerne
nucleon Nukleon
nucleus Atomkern, Kern
nuclide Nuklid
null vector [M] Nullvektor
number [M] Zahl
number Zahlenwert
number of loops Windungszahl
numerator [M] Zähler
nut Mutter
nut (guitar) Sattel (Gitarre)
nylon Nylon
oak wood Eichenholz
oar Ruder
object Gegenstand, Körper, Objekt
object distance Gegenstandsweite
object size Gegenstandsgröße
objective Objektiv
oblique throw [A. E.] schiefer Wurf, schräger Wurf
observation Beobachtung
octane [C] Octan
octave Oktave
odometer [A. E.] Kilometerzähler
ohm [unit] Ohm
Ohm's law Ohm'sches Gesetz

ohmic resistance ohmscher Widerstand (die Eigenschaft)
ohmic resistor ohmscher Widerstand (das elektrische Bauteil)
ohmmeter Ohmmeter
olive oil Olivenöl
Oort cloud Oortsche Kometenwolke
opal Opal
open electric circuit offener Stromkreis
open system offenes System
opposite [M] Gegenkathete
opposite in phase gegenphasig
opposite sign entgegengesetztes Vorzeichen
optical axis optische Achse
optical bench optische Bank
optical density optische Dichte (Extinktion)
optical depth optische Dicke, optische Tiefe
optical instrument optisches Instrument
optical power Brechkraft
optical thickness optische Dicke, optische Tiefe
optics Optik
to optimise optimieren
orbit Umlaufbahn
to orbit (a planet) umkreisen (einen Planeten)
orbit period Umlaufdauer
orbital radius Bahnradius
order of magnitude Größenordnung
order of the main maximum Ordnung des Hauptmaximums
order of the maximum Ordnung des Maximums
ore Erz
origin [M] Nullpunkt
O-ring O-Ring
oscillation Schwingung
oscillator velocity Oszillatorengeschwindigkeit, Schnelle (Oszillatorengeschwindigkeit)
osmium [Os, Z=76] Osmium
ounce [unit] Unze
outer atmosphere äußere Atmosphäre
outer ear Außenohr
outer space ferner Weltraum, fernes Weltall
outlet Steckdose
outwards [adj.] [adv.] nach außen
oval window (ear) ovales Fenster (Ohr)
oven Ofen
a over b [M] a geteilt durch b
overpressure Überdruck
overtone Oberschwingung, Oberton
to oxidise oxidieren
oxidised oxidiert
oxygen [O, Z=8] Sauerstoff
pain limit Schmerzgrenze
paintbox Malkasten

pair of compasses [rare] Zirkel
palladium [Pd, Z=46] Palladium
paperclip Büroklammer
parabola [M] Parabel
parabolic dish Parabolspiegel
parabolic mirror Parabolspiegel
parachute Fallschirm
paraffin Paraffin
paraffin oil [C] Paraffinöl
in parallel in Parallel
parallel [adj.] [adv.] parallel
parallel circuit Parallelschaltung
parallel connection Parallelschaltung
parallel plate capacitor Plattenkondensator
parallel ray Parallelstrahl
parallelogram [M] Parallelogramm
paramagnet Paramagnet
paramagnetic paramagnetisch
paramagnetic substance paramagnetische Substanz
paramagnetism Paramagnetismus
parent nucleus Mutterkern
part Bauteil, Teil
nth part [M] n-ter Teil
partial solar eclipse partielle Sonnenfinsternis
particle Teilchen
particle accelerator Teilchenbeschleuniger
particle accelerator ring Teilchenbeschleunigungsring
particle model Teilchenmodell
particle number Teilchenzahl
particle number density Teilchenzahldichte
particle track Teilchenspur
pascal [unit] Pascal
pascal second [unit] Pascalsekunde
Pascal's law Gesetz von Pascal
path Bahn, Flugbahn
path difference Gangunterschied, Pfaddifferenz
payload Nutzlast
PE (potential energy) potenzielle Energie
pendulum Pendel
pendulum bob Pendelkörper
pendulum clock Pendeluhr
pentane [C] Pentan
penumbra Halbschatten
to perform work Arbeit verrichten
perimeter [M] Umfang
period Schwingungsdauer
periodic periodisch
periodic system of the elements [C] Periodensystem
permanent magnet Permanentmagnet
permanent magnetism Permanentmagnetismus

permeability of vacuum magnetische Feldkonstante
permittivity of free space Dielektrizitätskonstante des Vakuums [alt], elektrische Feldkonstante, Permittivität des Vakuums [selten]
permittivity of vacuum Dielektrizitätskonstante des Vakuums [alt], elektrische Feldkonstante, Permittivität des Vakuums [selten]
perpendicular senkrecht [adj.], senkrecht [adv.]
perpendicularly senkrecht [adv.]
perpetuum mobile Perpetuum mobile
perpetuum motion machine Perpetuum mobile
petrol [B. E.] Benzin
petrol gauge Tankuhr
petroleum Petroleum
phase Aggregatzustand, Phase
phase (of a wave) Phase (einer Welle)
phase (of an oscillation) Phase (einer Schwingung)
phase (the phase wire) Phase (das elektrische Kabel)
phase difference Phasendifferenz
phase velocity Phasengeschwindigkeit, Vektor der Phasengeschwindigkeit
phase wire Phase (das elektrische Kabel)
phase-dependent amplitude phasenabhängige Amplitude
phase-shifted phasenverschoben
phenomena Phänomene
phenomenon Phänomen
Phillips screwdriver Kreuzschraubendreher
phosphorescence Nachleuchten, Phosphoreszenz
phosphorus [P, Z=15] Phosphor
photographic plate Fotoplatte
photon Photon
photon energy Photonenenergie
physical constant physikalische Konstante
physics formula physikalische Formel
pi bond [C] Pi-Bindung
piano string Klaviersaite
pie chart Kreisdiagramm, Kuchendiagramm
pin Bolzen (ohne Gewinde)
pin hole Nadelloch, Stiftloch
pincers Beißzange
ping-pong ball Tischtennisball
pinhole Nadelloch, Stiftloch
pinhole camera Lochkamera
pint [unit] Pint
pipe Rohr, Schlauch
pipe organ Pfeifenorgel
piston Kolben
piston rod Pleuel, Pleuelstange

pitch Tonhöhe
pitchblende Pechblende
plan Planzeichnung, Zeichnung
Planck constant Planck'sches Wirkungsquantum
planed gehobelt
planet Planet
planetary motion Planetenbewegung
planetary orbit Planetenbahn
planetoid Planetoid
plastic Plastik
plastic deformation plastische Deformation
plasticine Knete, Knetmasse
plate area Plattenfläche
plate distance Plattenabstand
platinum [Pt, Z=78] Platin
Plexiglas Plexiglas
plexiglass Plexiglas
pliers Zange
to plot [M] graphisch darstellen
plot [A. E.] [M] Diagramm
plug Stecker, Stopfen
plus [M] plus
a plus b [M] a plus b
plus pole Pluspol
plutonium [Pu, Z=94] Plutonium
pneumatic pneumatisch
pneumatic hammer Presslufthammer
pneumatic-tyred luftbereift
to point zeigen
point charge Punktladung
point of maximum height Punkt der maximalen Höhe
point source punktförmige Quelle
pointed spitz
point-like punktförmig
polar polar
polarisation Polarisation
polarity Polarität
pole Pol, Stab, Stange
to polish polieren
polished poliert
polonium [Po, Z=84] Polonium
polynomial [M] Polynom
polystyrene Styropor
porcelain Porzellan
position Ort
position vector Ortsvektor
positive pole positiver Pol
positive sign [M] positives Vorzeichen
positively charged positiv geladen
postulates of relativity Postulate der Relativitätstheorie
potassium [K, Z=19] Kalium
potential divider Spannungsteiler
potential energy potenzielle Energie

potentiometer Potenziometer
pound [unit] Pfund
pound-force [unit] Pound-force (US-amerikanische Krafteinheit)
pound-force per square inch [unit] psi (US-amerikanische Druckeinheit)
x to the nth power [M] x hoch n
power Leistung
power [M] Potenz
power (stroke 3 of a four-stroke engine) Arbeiten (Takt 3 eines Viertaktmotors)
power at a resistor Leistung in einem Widerstand
power generator Kraftwerksgenerator
power in a cylinder Zylinderleistung
power loss Leistungsverlust
power of a wave Leistung einer Welle
x to the power of n [M] x hoch n
power of ten [M] Zehnerpotenz
nth power of x [M] n-te Potenz von x
power outlet Steckdose
power plant Kraftwerk
power point Steckdose
power station Kraftwerk
praseodymium [Pr, Z=59] Praseodym
precise genau
prefix Präfix, Vorsilbe
prefixes Präfixe
pre-resistor Vorwiderstand
pressure Druck
pressure wave Druckwelle
presumption Vermutung
primary circuit Primärkreis
primary coil Primärspule
primary colour Grundfarbe
primary electron primäres Elektron
primary particle Primärteilchen
primary wave Primärwelle
principal axis optische Achse
principal focus Brennpunkt
principle of relativity Relativitätsprinzip
principle of the reversibility of the light path Prinzip der Umkehrbarkeit des Lichtwegs
printing method Druckverfahren
prism Prisma
to proceed straight geradeaus verlaufen
process Vorgang
product [M] Produkt
projectile motion schiefer Wurf, schräger Wurf
projection Projektion
promethium [Pm, Z=61] Promethium
propagation (of a wave) Ausbreitung (einer Welle)
propagation speed Fortpflanzungsgeschwindigkeit

propane [C] Propan
proper length Eigenlänge
proper time Eigenzeit
property Eigenschaft
property of a material Materialeigenschaft
proportional proportional
proportional to proportional zu
proportionality Proportionalität
proportionality constant Proportionalitätskonstante
protactinium [Pa, Z=91] Protactinium
proton Proton
proton number Protonenzahl
prototype kilogram Urkilogramm
prototype metre bar Urmeter
to provide power Leistung zur Verfügung stellen
pseudo force Scheinkraft
psi [unit] psi (US-amerikanische Druckeinheit)
pulley Riemenscheibe, Rolle, Umlenkrolle
pulling force Zugkraft
pump Pumpe
punched disc Lochscheibe
quantum Quant
quantum leap Quantensprung
quantum mechanics Quantenmechanik
quantum physics Quantenphysik
quart [unit] Quart
quartz Quarz
quartz glass Quarzglas
quartz watch Quarzuhr
to question hinterfragen
quotient [M] Produkt, Quotient
radial radial
radially inwards [adv.] radial nach innen
radially outwards [adv.] radial nach außen
radian [M] Radian
radian measure [M] Bogenmaß
radiance (power per area per solid angle) Strahldichte (Leistung pro Fläche pro Raumwinkel)
radiation radioaktive Strahlung, Strahlung
radiation belt Strahlungsgürtel
radiation shielding Abschirmung radioaktiver Strahlung
radiative transfer Strahlungsdurchgang
radiator Heizkörper
radicand [M] Radikand
radioactive radioaktiv
radioactive decay radioaktiver Zerfall
radioactive decay law radioaktives Zerfallsgesetz
radioactive isotope radioaktives Isotop
radioactivity Radioaktivität
radiocarbon dating Radiokarbonmethode
radium [Ra, Z=88] Radium
radius [M] Radius

radon [Rn, Z=86] Radon
rail Schiene
railway line Bahnlinie, Eisenbahnstrecke
railway track Gleisanlage, Schiene
rape seed oil Rapssamenöl
rarefaction disturbance Verdünnungsstörung
rational number [M] rationale Zahl
RCD (residual current device) Fehlerstromschutzschalter, FI-Schutzschalter, FI-Sicherung
reaction time Reaktionszeit
to readjust neu justieren
real current direction physikalische Stromrichtung
real function [M] reelle Funktion
real image reelles Bild
real number [M] reelle Zahl
to rebound abprallen
reciprocal [M] Kehrwert
reciprocally proportional umgekehrt proportional
reciprocally proportional to umgekehrt proportional zu
recoil Rückstoß
to record aufzeichnen
rectangle [M] Rechteck
redshift Rotverschiebung
to reflect reflektieren
reflectance Reflektanz, Reflexionsgrad
reflected ray reflektierter Strahl
reflection Reflexion, Spiegelung
reflection angle Reflexionswinkel
reflectivity Reflektanz
refraction Brechung
refraction angle Brechungswinkel
refractive index Brechzahl, optische Dichte (Brechzahl)
regression line Regressionsgerade
regular reflection regelmäßige Reflexion
relation Beziehung
relative permeability Permeabilitätszahl
relative permittivity Dielektrizitätszahl
relativistic energy-momentum relation relativistische Energie-Impuls-Beziehung
relativistic kinetic energy relativistische kinetische Energie
relativistic mass relativistische Masse
relativistic momentum relativistischer Impuls
relativity Relativität
relativity principle Relativitätsprinzip
to release power Leistung freisetzen
remnant (of a star) Überrest (eines Sterns)
to repel abstoßen
to repel each other sich gegenseitig abstoßen
representation Darstellung

repulsion Abstoßung
repulsive abstoßend
to re-scale (a vector) [M] skalieren (einen Vektor)
residual current device Fehlerstromschutzschalter, FI-Schutzschalter, FI-Sicherung
resistance Widerstand (physikalische Größe)
resistance thermometer Widerstandsthermometer
resistivity spezifischer Widerstand
resistor Widerstand (Bauteil)
resistor combination Widerstandskombination
resolution of a vector [M] Vektorzerlegung
to resolve (a vector) [M] zerlegen (einen Vektor)
resonance Resonanz
resonance disaster Resonanzkatastrophe
resonant frequency Resonanzfrequenz
resonant value of the amplitude Resonanzamplitude
to rest ruhen
rest energy Ruheenergie
rest frame Ruhesystem
rest mass Ruhemasse
rest position Ruhelage
restoring force rücktreibende Kraft
resultant force resultierende Kraft
resulting oscillation resultierende Schwingung
reticle Fadenkreuz
retina Netzhaut, Retina
return spring Rückholfeder
rev counter Drehzahlmesser
reversibility of the light path Umkehrbarkeit des Lichtwegs
revolution counter Drehzahlmesser
rhenium [Re, Z=75] Rhenium
rhodium [Rh, Z=45] Rhodium
ribbon Band
ribbon generator Bandgenerator
right hand rule Rechte-Hand-Regel
right triangle [A. E.] [M] rechtwinkliges Dreieck
right-angled triangle [B. E.] [M] rechtwinkliges Dreieck
rigid steif
rim Felge
ripple tank Wellenwanne
to rise (e. g. in water) aufsteigen (z. B. in Wasser)
rising time Aufstiegszeit, Steigzeit
robot Roboter
rocket Rakete (Raumfahrt)
rocket stage Raketenstufe
rod Stab, Stange
rods (eye) Stäbchen (Auge)

roller Walze
roller coaster Achterbahn
rolling friction Rollreibung
rolling friction coefficient Rollreibungszahl
rolling friction force Rollreibungskraft
root [M] Wurzel
nth root of x [M] n-te Wurzel von x
rope Seil
rotary table Drehtisch
rotation Drehbewegung, Drehung, Rotation
rotor Rotor
rough surface raue Oberfläche
roughness Rauheit
to rub reiben
rubber Gummi
rubber band Gummiband
rubidium [Rb, Z=37] Rubidium
rule of three [M] Dreisatz
rule of thumb Daumenregel
ruler Lineal
rules Regeln
runway Landebahn, Startbahn
ruthenium [Ru, Z=44] Ruthenium
rutherfordium [Rf, Z=104] Rutherfordium
safety fuse Sicherung
sailplane Segelflugzeug
salt [C] Salz
salt crystal [C] Salzkristall
salt water Salzwasser
samarium [Sm, Z=62] Samarium
same sign dasselbe Vorzeichen
same size image Bild gleicher Größe
sand Sand
sandpaper Schmirgelpapier
sapphire Saphir
satellite Satellit
satellite dish Satellitenschüssel
Saturn Saturn
saw Säge
to saw sägen
sawdust Sägemehl
scalar Skalar
scalar multiplication [M] Skalarmultiplikation
scale Skala, Waage
scandium [Sc, Z=21] Scandium
scattering Streuung
Schrödinger equation Schrödingergleichung
science Wissenschaft
scientific notation wissenschaftliche Schreibweise
scissors Schere
screen Schirm
screw Holzschraube
screwdriver Schraubendreher
to scrutinise hinterfragen

sea level Meereshöhe
sea water Meerwasser, Salzwasser
seaborgium [Sg, Z=106] Seaborgium
second [unit] Sekunde
second Kepler law zweites Kepler'sches Gesetz
second law of thermodynamics zweiter Hauptsatz der Thermodynamik
second Newtonian law zweites Newton'sches Gesetz
secondary circuit Sekundärkreis
secondary coil Sekundärspule
secondary colour Mischfarbe
secondary electron sekundäres Elektron
secondary particle Sekundärteilchen
secondary rainbow sekundärer Regenbogen
secondary wave Sekundärwelle
seismic wave seismische Welle
seismogram Seismogramm
selenium [Se, Z=34] Selen
semicircle [M] Halbkreis
semicircular [M] halbkreisförmig
semicircular canals (ear) Bogengänge (Ohr)
semiconductor Halbleiter
semimajor axis große Bahnhalbachse
semitone Halbton
semolina grain Grieskorn
sensory cells Sinneszellen
in series in Reihe
series circuit Reihenschaltung
series connection Reihenschaltung
set [M] Menge
setup Aufbau
set-up Aufbau
shadow Schatten
shadow formation equation Schattenbildungsgleichung
shadow space Schattenraum
shallow [M] flach
shape Form
sharp scharf, spitz
sharpness Schärfe
shear force Scherkraft
shear wave Scherwelle
to shift [M] verschieben
shifted [M] verschoben
to shine (e. g. a light bulb) leuchten (z. B. eine Glühbirne)
shiny glänzend
SHM (simple harmonic motion) harmonische Schwingung
shock absorber Stoßdämpfer
short Kurzschluss
short circuit Kurzschluss
to short-circuit kurzschließen
short-circuit Kurzschluss

to shorten a vector [M] einen Vektor stauchen
short-sighted kurzsichtig
short-sightedness Kurzsichtigkeit
SI-base quantity SI-Basisgröße
SI-base unit SI-Grundeinheit
sievert [unit] Sievert
sigma bond [C] Sigma-Bindung
sign [M] Vorzeichen
significant figures signifikante Stellen
silicon [Si, Z=14] Silizium
silicone oil Silikonöl
silver [Ag, Z=47] Silber
simple harmonic motion harmonische Schwingung
simple pendulum Fadenpendel
sine [M] Sinus
sine curve [M] Sinuskurve
single slit Einfachspalt
to sink (e. g. in water) untergehen (z. B. im Wasser)
sinus [M] Sinus
sinusoid [M] Sinusfunktion, Sinuskurve
sinusoidal [M] sinusförmig
siren Sirene
siren disc Lochsirene
SI-unit SI-Einheit
sketch Skizze
ski jumper Skispringer
skid marks Bremsspur
skin resistance Übergangswiderstand der Haut
slide Dia
to slide gleiten
slider (on an optical bench) Reiter (auf einer optischen Bank)
sliding contact Gleitkontakt
slipstream Windschatten
slit aperture Spaltblende
slope [M] Steigung
slope (of an inclined plane) Steigung (einer schiefen Ebene)
slope angle (of an inclined plane) Steigungswinkel (einer schiefen Ebene)
smoke detector Rauchmelder
smooth surface glatte Oberfläche
socket Steckdose
sodium [Na, Z=11] Natrium
sodium chloride [C] Natriumchlorid
sodium lamp Natriumdampflampe
to soften (sharpness) sich abschwächen (Schärfe)
solar distance (of a planet) Sonnenabstand (eines Planeten)
solar power station Solarkraftwerk
solar wind Sonnenwind
to solder löten

soldering iron Lötkolben
solenoid Spule (sehr groß)
solid fest
solid angle [M] Raumwinkel
sonar Echolot
sonic Schall-
sonic sender Schallsender
soot Ruß
sound Schall
sound generator Schallerzeuger, Tonerzeuger
sound level Schallpegel
sound pressure Schalldruck
sound pressure level Schalldruckpegel, Schallpegel
sound wave Schallwelle
South Pole Südpol (der Erde)
south pole Südpol (eines Magneten)
southern lights südliches Polarlicht, Südlicht
soy been oil Sojabohnenöl
soya been oil Sojabohnenöl
space Weltall, Weltraum
space flight Raumfahrt
spacecraft Raumschiff
spaceship Raumschiff
spacial [M] räumlich
spare part Ersatzteil
spark plug Zündkerze
special case Spezialfall
special theory of relativity spezielle Relativitätstheorie
specific heat spezifische Wärme
specific heat capacity spezifische Wärmekapazität
specific latent heat of fusion spezifische Schmelzwärme
specific latent heat of vaporization spezifische Verdampfungswärme
specific resistance spezifischer Widerstand
spectra Spektren
spectral colours Spektralfarben
spectroscope Spektroskop
spectrum Spektrum
spectrum of visible light Spektrum des sichtbaren Lichts
specular reflection regelmäßige Reflexion
speed Geschwindigkeit
speed amplitude Geschwindigkeitsamplitude
speed of light in a medium Lichtgeschwindigkeit in einem Medium
speed of light in vacuum [unit] Lichtgeschwindigkeit im Vakuum
speedometer Geschwindigkeitsmesser
sphere Kugel
sphere [M] Kugel
spheric wave Kugelwelle

spiky spitz
to spin um die eigene Achse rotieren
spiral Spirale
SPL (sound pressure level) Schalldruckpegel, Schallpegel
spreadsheet Messtabelle
spring Feder
to spring federn
spring constant Federkonstante
spring energy Federenergie
spring force Federkraft
spring pendulum Federpendel
spring pendulum mechanism Federpendelmechanismus
spring scale Federwaage
spruce wood Fichtenholz
square [M] Quadrat
square metre [unit] Quadratmeter
square root [M] Quadratwurzel
stand Stativ
standard acceleration due to gravity Normwert der Fallbeschleunigung
standard conditions Normbedingungen
standard pressure Normdruck
standard state Normzustand
standard temperature Normtemperatur
standard value Normwert
standing wave stehende Welle
start of a vector [M] Anfangspunkt eines Vektors
starting point Anfangspunkt
state Zustand
state of motion Bewegungszustand
state variables Zustandsgrößen
static friction Haftreibung
static friction coefficient Haftreibungszahl
static friction force Haftreibungskraft
stationary wave stehende Welle
stator Stator
steam engine Dampfmaschine
steel Stahl
steel cable Drahtseil, Stahlseil
steel rope Drahtseil, Stahlseil
steep [M] steil
Stefan-Boltzmann constant Stefan-Boltzmann-Konstante
Stefan-Boltzmann law Stefan-Boltzmann-Gesetz
stirrup Steigbügel
stirrup (ear) Steigbügel (Ohr)
Stokes' law Stokes-Gesetz
stop position Halteposition, Ruheposition
stop watch Stoppuhr
stove Herd
straight line [M] Gerade
straight line through zero [M] Nullpunktgerade
straight wave gerade Welle
strain relative Längenänderung
straw Strohhalm
streamline Stromlinie
stress Spannung
to stretch a vector [M] strecken (einen Vektor)
string Saite
string instrument Saiteninstrument, Streichinstrument
string pendulum Fadenpendel
stringed instrument Streichinstrument
stroboscope Stroboskop
stroke Takt (Motor)
strong force starke Kraft
strong interaction starke Wechselwirkung
strong nuclear force starke Kernkraft
strontium [Sr, Z=38] Strontium
structured aufgebaut, strukturiert
styrofoam Styropor
to sublimate sublimieren
sublimation Sublimation
subset of [M] Teilmenge von
substance Substanz
substrate Untergrund, Unterlage
substratum Untergrund, Unterlage
to subtract [M] subtrahieren
subtraction [M] Subtraktion
subtractive mixture subtraktive Mischung
subtrahend [M] Subtrahend
sulfur dioxide [C] Schwefeldioxid
sulfuric acid [C] Schwefelsäure
sulphur [S, Z=16] Schwefel
sum [M] Summe
sun Sonne
superconductor Supraleiter
supernova Supernova
superposition Superposition, Überlagerung
supersaturated übersättigt
supersaturated vapour übersättigter Dampf
supersaturated water vapour übersättigter Wasserdampf
surface area [M] Oberfläche
surface gravity Fallbeschleunigung an der Oberfläche
surface temperature Oberflächentemperatur
surface tension Oberflächenspannung
suspension Federung
to swim schwimmen (Lebewesen auf dem Wasser durch eigenen Krafteinsatz)
switch Schalter
symbol Symbol
synchronous motor Synchronmotor
syringe Spritze
system System

table Messtabelle
table of nuclides Nuklidtafel
table salt [C] Kochsalz
table salt crystal [C] Kochsalzkristall
tablespoon [unit] Esslöffel
to tabulate tabellieren
tabulated tabelliert
tachometer Drehzahlmesser
to take an X-ray of someone jemanden röntgen
tangent [M] Tangens
tangential [M] tangential [adj.]
tangentially [M] tangential [adv.]
tantalum [Ta, Z=73] Tantal
tap Wasserhahn
tar Teer
to tar teeren
to tare tarieren
target Ziel, Zielscheibe
tarmac Asphalt
to tarmac asphaltieren
teaspoon [unit] Teelöffel
technetium [Tc, Z=43] Technetium
technical application technische Anwendung
technique Technik
Teflon Teflon
telescope Fernrohr
tellurium [Te, Z=52] Tellur
temperature Temperatur
temperature of the mixture Mischtemperatur
temperature sensor Temperaturfühler
template Muster
tensile strain relative Längenänderung
tensile strength Zugfestigkeit
tensile stress Spannung
tension Spannkraft
terbium [Tb, Z=65] Terbium
terrestrial radiation Bodenstrahlung
tesla [unit] Tesla
test charge Probeladung
test tube Reagenzglas
textiles Textilien
thallium [Tl, Z=81] Thallium
theory Theorie
theory of relativity Relativitätstheorie
thermal electrons thermische Elektronen
thermal expansion Wärmeausdehnung
thermal expansion coefficient Wärmeausdehnungskoeffizient
thermal imaging camera Wärmebildkamera
thermal radiation Wärmestrahlung
thermochrome lacquer Thermochromlack
thermochrome paper Thermochrompapier
thermodynamic data thermodynamische Daten
thermodynamic equilibrium thermodynamisches Gleichgewicht
thermodynamics Thermodynamik, Wärmelehre
thermography Thermographie
thermometer Thermometer
thermos flask Thermosflasche
third Kepler law drittes Kepler'sches Gesetz
third law of thermodynamics dritter Hauptsatz der Thermodynamik
third Newtonian law drittes Newton'sches Gesetz
thorium [Th, Z=90] Thorium
thread Gewinde
three finger rule Dreifingerregel
three finger rule of the right hand Dreifingerregel der rechten Hand
three wire cable dreiadriges Kabel
throwing distance Wurfweite
thulium [Tm, Z=69] Thulium
tidal Gezeiten-
tides Gezeiten
timbre Klang
time Zeit
time constant Zeitkonstante
time dilation Zeitdilatation
time period Umlaufdauer
times [M] mal
a times b [M] a mal b
timetable Fahrplan
tin [Sn, Z=50] Zinn
tip of a vector [M] Endpunkt eines Vektors, Spitze eines Vektors
titanium [Ti, Z=22] Titan
toggle Umschalter
toluol [C] Toluol
tone Ton
tone pitch Tonhöhe
toner cartridge Tonerkartusche, Tonerpatrone
torch Taschenlampe
torque Drehmoment
torque wrench Drehmomentschlüssel
total energy Gesamtenergie
total internal reflection Totalreflexion
total pressure Gesamtdruck
total reflection Totalreflexion
track Bahngleis
train carriage Eisenbahnwagen
train wagon Eisenbahnwagon
trajectory Trajektorie
tram Tram
trampoline Trampolin
to transfer übertragen
transformer Transformator

transformer current relation Stromstärkebeziehung beim Transformator
transformer equation Transformatorengleichung
transformer power relation Leistungsbeziehung beim Transformator
transformer station Transformatorenstation
transformer voltage relation Spannungsbeziehung beim Transformator
transition Übergang
transmission line Hochspannungsleitung
to transmit übermitteln
transmittance Transmissionsgrad
transverse coupling transversale Kopplung
transverse wave transversale Welle
triangle [M] Dreieck
trigonometry [M] Trigonometrie
trough Wellental
troy ounce [unit] Feinunze
troy pound [unit] Apotheker-Pfund
tube Rohr, Schlauch
tug Schleppschiff
tugboat Schleppschiff
tungsten [W, Z=74] Wolfram
tuning fork Stimmgabel
tuning key (guitar) Wirbel (Gitarre)
turbulence Turbulenz
turbulent flow turbulente Strömung
turquoise türkis
two-dimensional motion zweidimensionale Bewegung
two-stroke engine Zweitaktmotor
types of waves Wellenarten
tyre Reifen
tyre pump Luftpumpe
ultimate tensile strength Zugfestigkeit
ultrasonic Ultraschall-
ultrasound Ultraschall
ultraviolet light ultraviolettes Licht
umbra Kernschatten
unambiguous eindeutig
uncalibrated unkalibriert
uncombined [C] ungebunden
undamped oscillation ungedämpfte Schwingung
underpressure [rare] Unterdruck
undesirable friction unerwünschte Reibung
uniform circular motion gleichförmige Kreisbewegung
uniform motion gleichförmige Bewegung
uniform motion with starting point gleichförmige Bewegung mit Anfangspunkt
uniformly accelerated motion gleichmäßig beschleunigte Bewegung
uniformly accelerated motion with initial speed gleichmäßig beschleunigte Bewegung mit Anfangsgeschwindigkeit
uniformly accelerated motion with initial speed and with starting point gleichmäßig beschleunigte Bewegung mit Anfangsgeschwindigkeit und Anfangspunkt
unique eindeutig
unit Einheit
unit circle [M] Einheitskreis
universal generator Universalgenerator
universal motor Universalmotor
universe Universum
unlike poles ungleichnamige Pole
unstable instabil
upright image aufrechtes Bild
upthrust [rare] Auftrieb, Auftriebskraft
uranium [U, Z=92] Uran
uranium ore Uranerz
uranium rock Urangestein
uranium-lead method Uran-Blei-Methode
Uranus Uranus
to use energy Energie verbrauchen
U-tube U-Rohr
U-tube manometer U-Manometer
vacuum Unterdruck, Vakuum
vacuum cleaner Staubsauger
vacuum tube Vakuumröhre
value Wert
valve Ventil
Van Allen belt Van-Allen-Gürtel
van de Graaff generator Van de Graaff Generator
vanadium [V, Z=23] Vanadium
vaporisation Verdampfen
to vaporise verdampfen
vapour Dampf
variable Variable
variable capacitor Drehkondensator
variable resistor Schiebewiderstand
variation of A with B Abhängigkeit von A von B
V-belt Keilriemen
vector [M] Vektor
vector addition [M] Vektoraddition
vector analysis [M] Vektoranalysis
vector decomposition [M] Vektorzerlegung
velocity Geschwindigkeitsvektor
Venus Venus
vertical senkrecht [adj.]
vertical throw vertikaler Wurf
vertically senkrecht [adv.]
vibration generator Schwingungsgenerator
vibratory system schwingungsfähiges System
vice Schraubstock
virtual focal point virtueller Brennpunkt
virtual image virtuelles Bild
viscosity Viskosität

viscous damping coefficient viskoser Dämpfungsfaktor
visibility Sichtweite
visible light sichtbares Licht
to visualise (field lines) sichtbar machen (Feldlinien)
vocal cord Stimmband
volt [unit] Volt
voltage Spannung
voltage divider Spannungsteiler
voltage drop Spannungsabfall
voltage drops at a resistor Spannung fällt an einem Widerstand ab
voltage source Spannungsquelle
voltages in the parallel circuit Spannungen bei der Parallelschaltung
voltages in the series circuit Spannungen bei der Reihenschaltung
voltmeter Voltmeter
volume Volumen
volumetric mean radius Radius der volumengleichen Kugel
volumetric thermal expansion Volumenausdehnung
volumetric thermal expansion coefficient Volumenausdehnungskoeffizient
wand Zauberstab
water displacement Wasserverdrängung
water jet Wasserstrahl
water reservoir Wasserreservoir
water vapour Wasserdampf
water wave Wasserwelle
watt [unit] Watt
wave Welle
wave crest Wellenberg
wave equation Wellengleichung
wave front Wellenfront
wave number Wellenzahl
wave trough Wellental
wavelength Wellenlänge
wavenumber Wellenzahl
wax ball Wachskugel
weak force schwache Kraft
weak interaction schwache Wechselwirkung
weak nuclear force schwache Kernkraft
wearing part Verschleißteil
weber [unit] Weber
Wehnelt tube Wehneltröhre
weight Gewicht
weightlessness Schwerelosigkeit
to weld schweißen
welder Schweißgerät
wheel rim Felge
wheels lock up Reifen blockieren
wheels spin Räder drehen durch
white dwarf weißer Zwerg

integer [M] ganze Zahl
width Breite
winch Seilwinde, Winde
wind instrument Blasinstrument
wind power station Windkraftwerk
to wind up something etwas aufwickeln
wind-driven power station Windkraftwerk
window glass Fensterglas
window pane Fensterscheibe
wing (aeroplane) Tragfläche (Flugzeug)
wire Draht
wire loop Leiterschlaufe
wood Holz
work Arbeit
work in a cylinder Zylinderarbeit
work performed on a spring Federarbeit
world view Weltbild
worst case scenario GAU, größter annehmbarer Unfall
wound capacitor Wickelkondensator
wrench Schraubenschlüssel
wrist watch Armbanduhr
xenon [Xe, Z=54] Xenon
to X-ray röntgen
X-ray Röntgenstrahl
X-ray photon Röntgenphoton
X-ray quantum Röntgenquant
yard [unit] Yard
yellow gelb, yellow
y-intercept [M] y-Achsenabschnitt
ytterbium [Yb, Z=70] Ytterbium
yttrium [Y, Z=39] Yttrium
zero gravity Schwerelosigkeit
zero point [M] Nullpunkt
zero-crossing Nulldurchgang
zinc [Zn, Z=30] Zink
zirconium [Zr, Z=40] Zirkon

Teil 2
Deutsch - Englisch

1. **Hauptsatz der Thermodynamik** 1st law of thermodynamics
1. **Kepler'sches Gesetz** 1st Kepler law
1. **Newton'sches Gesetz** 1st Newtonian law
2. **Hauptsatz der Thermodynamik** 2nd law of thermodynamics
2. **Kepler'sches Gesetz** 2nd Kepler law
2. **Newton'sches Gesetz** 2nd Newtonian law
3. **Hauptsatz der Thermodynamik** 3rd law of thermodynamics
3. **Kepler'sches Gesetz** 3rd Kepler law
3. **Newton'sches Gesetz** 3rd Newtonian law
abbilden [M] to map
Abbildung image formation
Abbildung [M] mapping
Abbildungsgleichung image formation equation
Abfall decrease
abgeleitete SI-Einheit derived SI-Unit
abgeschlossenes System closed system
Abhängigkeit von A von B dependence of A on B, variation of A with B
abklingen to attenuate, to decay (oscillation)
Ablation ablation
ableiten to derive
ableiten [M] to derive
ableiten (logisch) to deduce
Ableitung [M] derivation
Abnahme decrease
Abnutzung deterioration
abprallen to rebound
abschätzen to estimate
Abschätzung estimation
Abschirmung radioaktiver Strahlung radiation shielding
sich abschwächen to attenuate, to decay (oscillation)
sich abschwächen (Schärfe) to soften (sharpness)
absolute Temperatur absolute temperature
absoluter Nullpunkt absolute zero
Absorption absorption
Absorptionsgrad absorbance
Abstandsvektor displacement
abstoßen to repel
abstoßend repulsive
Abstoßung repulsion
Aceton [C] acetone
Acetylen [C] acetylene
x-Achse [M] x-axis
y-Achse [M] y-axis
Achse [M] axis
Achse axle
Achse des magnetischen Feldes magnetic field axis
Achterbahn roller coaster

Actinium actinium [Ac, Z=89]
Actio=Reactio actio=reactio
addieren [M] to add
Addition [M] addition
additive Mischung additive mixture
Adiabatenkoeffizient adiabatic coefficient
Aerometer aerometer
Aggregatzustand phase
Akkommodation accommodation
Aktivität activity
Aktivitätsgesetz activity law
Akustik acoustics
Algebra [M] algebra
allgemeine Gasgleichung ideal gas equation
allgemeine Gaskonstante gas constant
allgemeine Relativitätstheorie general theory of relativity
Alpharadioaktivität alpha radioactivity
Alphateilchen alpha particle
Alphazerfall alpha decay
altern to age
Altern aging
aluminiert aluminized
Aluminium aluminium [Al, Z=13]
Amboss (Ohr) anvil (ear)
Amboss (Werkzeug) anvil (tool)
Americium americium [Am, Z=95]
Ammoniak [C] ammonia
Amonton'sches Gesetz Amonton's law
Ampere [Einh.] amp, ampere
Amperemeter ammeter, amperemeter
Amplitude amplitude
Analyse analysis
Analysis [M] analysis
anfängliche Aktivität initial activity
Anfangsgeschwindigkeit initial speed
Anfangspunkt starting point
Anfangspunkt eines Vektors [M] initial point of a vector, start of a vector
anfertigen to manufacture
Anfertigung manufacture
angeregt (Atom, Elektron, Atomkern) excited (atom, electron, nucleus)
angeregter Zustand excited state
angetriebene Achse driven axle
Angström [Einh.] angstrom
Anion [C] anion
Ankathete [M] adjacent
Anode anode
Anpassungskurve fit
anregen (Atom, Elektron, Atomkern) to excite (atom, electron, nucleus)
anregen (Schwingung) to excite (oscillation)
anregen (Welle) to excite (wave)
Anregung (Atom, Elektron, Atomkern) excitation (atom, electron, nucleus)

Ansaugen (Takt 1 eines Viertaktmotors) intake (stroke 1 of a four-stroke engine)
Anschlag (z. B. einer Federwaage) end stop (e. g. of a spring scale)
Anstieg increase
Antimon antimony [Sb, Z=51]
Antriebsriemen drive belt
Antriebsriemenscheibe drive pulley
Antriebsscheibe drive pulley
Anwendung application
sich anziehen to attract
anziehen to attract
anziehend attractive
Anziehung attraction
Apertur aperture
Apotheker-Pfund [Einh.] troy pound
Aquarium aquarium
Äquatorradius equatorial radius
äquidistant equidistant
Äquivalentdosis equivalent dose
Äquivalenzdosis equivalent dose
Arbeit work
Arbeit des elektrischen Stroms electrical work
Arbeit verrichten to do work, to perform work
Arbeiten (Takt 3 eines Viertaktmotors) power (stroke 3 of a four-stroke engine)
arcus [M] arcus
Are [Einh.] are
Argon argon [Ar, Z=18]
Armbanduhr wrist watch
Arsen arsenic [As, Z=33]
Asphalt asphalt, tarmac
asphaltieren to tarmac
Assimilation assimilation
Astat astatine [At, Z=85]
Asteroid asteroid
Asteroidengürtel asteroid belt
Astronaut astronaut
Astronomie astronomy
astronomische Einheit [Einh.] astronomical unit
Atmosphäre [Einh.] atmosphere
Atmosphäre (z. B. der Erde) atmosphere (e. g. of Earth)
Atom atom
atomare Ladungseinheit [Einh.] atomic unit of charge
atomare Masseneinheit atomic mass unit
Atombau atomic structure
Atomgitter atomic lattice
Atomhülle atomic shell
Atomkern atomic nucleus, nucleus
Atomkerne atomic nuclei, nuclei
Atomkraftwerk nuclear power plant
Atomschale atomic shell
Atomsprengkopf nuclear warhead
Atomstruktur atomic structure
Aufbau makeup, setup, set-up
aufblasen to inflate
aufgebaut structured
aufladen to charge
Aufladen charging
Aufladung charging, charging-up
aufrechtes Bild upright image
aufsteigen (z. B. in Wasser) to rise (e. g. in water)
Aufstiegszeit rising time
Auftrieb buoyancy, upthrust [rare]
Auftriebskraft buoyancy, upthrust [rare]
sich aufwickeln to coil itself up
aufzeichnen to record
Aufzug elevator [A. E.], lift [B. E.]
Aufzugkabine elevator cab [A. E.], elevator cabin [A. E.], lift car [B. E.]
Augapfel eyeball
Auge eye
Augenlinse eye lens
Ausbreitung (einer Welle) diffusion (of a wave), propagation (of a wave)
Ausdehnung extension
ausklammern [M] to factorize
Auslenkung deflection, elongation
ausrichten to align
Ausrüstung equipment
nach außen outwards [adj.] [adv.]
Außenohr outer ear
äußere Atmosphäre outer atmosphere
Äußeres exterior
Ausstoßen (Takt 4 eines Viertaktmotors) exhaust (stroke 4 of a four-stroke engine)
Ausstrahlung (Leistung pro Oberfläche) excitance (power per surface area)
auswerten to evaluate
Auswertung evaluation
Avogadro-Zahl Avogadro's number
Bagger excavator
Baggerschaufel excavator shovel
Bahn path
Bahngleis track
Bahnlinie railway line
Bahnradius orbital radius
Bananenstecker banana plug
Band ribbon
Bandgenerator ribbon generator
Bar [Einh.] bar
Barium barium [Ba, Z=56]
Barometer barometer
barometrische Höhenformel barometric formula
Barrel [Einh.] barrel

Basis (einer Potenz) [M] base (of a power)
Bassgitarre bass guitar
Basssaite bass string
Batterie battery
Baumwolltuch cotton cloth
Bauteil component, part
Becherglas beaker
Becquerel [Einh.] becquerel
Beißzange pincers
belasten to load
Beleuchtung illumination
Beleuchtungsstärke illuminance
bemannte Mission manned mission
Benzin gas [A. E.], gasoline [A. E.], petrol [B. E.]
Benzol [C] benzene
Beobachtung observation
Berkelium berkelium [Bk, Z=97]
Bernstein amber
Berührungsfläche contact area
Beryllium beryllium [Be, Z=4]
beschleunigen to accelerate
Beschleuniger accelerator
Beschleunigung acceleration
Beschleunigungsamplitude acceleration amplitude
Beschleunigungsarbeit acceleration work
Beschleunigungseinheit acceleration unit
Beschleunigungs-Elongationsbeziehung acceleration-elongation relation
Beta-Faktor beta factor
Betaradioaktivität beta radioactivity
Betateilchen beta particle
Betazerfall beta decay
Beton concrete
Betrag [M] norm
Betrag eines Vektors [M] magnitude of a vector, norm of a vector
Beugung diffraction
Beugungsgitter diffraction grating, grating
Beugungswinkel diffraction angle
Bewegung motion
Bewegungsgleichung equation of motion
Bewegungsrichtung direction of motion
Bewegungszustand state of motion
Beziehung relation
Bild image
Bild gleicher Größe same size image
Bildgröße image size
bildlich darstellen to depict
Bildorientierung image orientation
Bildweite image distance
Bimetallthermometer bimetallic thermometer
Bindung [C] bond
Bindungskräfte [C] binding forces

Biot-Savart-Kraft electromotive force
Birkenholz birch wood
Bismut bismuth [Bi, Z=83]
Blasinstrument wind instrument
Blasrohr blow pipe, blowgun
Blauverschiebung blueshift
Blei lead [Pb, Z=82]
Blende aperture
Blitz lightning
Blitzableiter lightning conductor
Bodenstrahlung terrestrial radiation
Bogengänge (Ohr) semicircular canals (ear)
Bogenlänge [M] arc length, length of arc
Bogenmaß [M] circular measure, radian measure
Bohrer drill
Bohrium bohrium [Bh, Z=107]
Bohrmaschine drill
Boltzmann-Konstante Boltzmann constant
Bolzen (mit Gewinde) bolt
Bolzen (ohne Gewinde) pin
Bor boron [B, Z=5]
Boyle-Mariotte'sches Gesetz Boyle-Mariotte's law
Boyle-Mariotte-Experiment Boyle-Mariotte experiment
Brechkraft optical power
Brechung refraction
Brechungsgesetz law of refraction
Brechungswinkel refraction angle
Brechzahl refractive index
Breite width
Breitengrad latitude
Bremsbelag brake pad
Bremse brake
bremsen to brake
Bremsen braking
Bremsflüssigkeit brake fluid
Bremsleitung brake line
Bremspedal brake pedal
Bremsscheibe brake disk
Bremsschlauch brake hose
Bremsschuh brake shoe
Bremsspur skid marks
Bremstrommel brake drum
Bremsverzögerung deceleration
Bremsweg braking distance
Bremszeit braking time
Bremszylinder brake cylinder
Brennebene focal plane
Brenner burner
Brennkammer combustion chamber
Brennpunkt focal point, focus, principal focus
Brennstrahl focal ray
Brennweite focal length

Brennwert calorific value, gross calorific value, gross heat of combustion, heat of combustion, heating value, higher heating value
Brille glasses
Brom bromine [Br, Z=35]
Bruch [M] fraction
Bruchstrich [M] fraction bar
Buchenholz beech wood
Bügeleisen iron
Bund (Gitarre) fret (guitar)
Bundstab (Gitarre) fret (guitar)
Büroklammer paperclip
Bürste brush
Bürste (Motor, Generator) brush (motor, generator)
Butan [C] butane
C-14-Anteil C-14 portion
C-14-Methode C-14 dating
Calcium calcium [Ca, Z=20]
Californium californium [Cf, Z=98]
Candela [Einh.] candela
Cäsium caesium [Cs, Z=55]
C-C-Bindung [C] carbon-carbon bond
Celsius-Temperatur Celsius temperature
Cer cerium [Ce, Z=58]
Chemie chemistry
chemische Bindung [C] chemical bond
Chladni'sche Klangfiguren Chladni patterns
Chlor chlorine [Cl, Z=17]
Chrom chromium [Cr, Z=24]
Corioliskraft Coriolis force
Cosinus [M] cosine, cosinus
Coulomb [Einh.] coulomb
Coulomb-Gesetz Coulomb's law
Coulombkraft Coulomb force
Curie [Einh.] curie
Curietemperatur Curie temperature
Curium curium [Cm, Z=96]
cyan cyan
Dampf vapour
Dampfmaschine steam engine
Dämpfung damping
Dämpfungsfaktor damping coefficient
darstellen to depict
Darstellung representation
dasselbe Vorzeichen same sign
Datenpunkt data point
Daumenregel rule of thumb
de Broglie-Beziehung de Broglie relation
Deckel lid
Deformation deformation
deformieren to deform
Deklination declination
destilliertes Wasser distilled water
destruktive Interferenz destructive interference

Dezimalpunkt [M] decimal point
Dezimalstelle [M] decimal place
Dia slide
Diagramm [M] chart, diagram, graph, plot [A. E.]
Diamagnet diamagnet
diamagnetisch diamagnetic
diamagnetische Substanz diamagnetic substance
Diamagnetismus diamagnetism
Diamant diamond
Dichte density
Dielektrika dielectrics
Dielektrikum dielectric
Dielektrizitätskonstante des Vakuums [alt] electric constant, electric field constant [rare], permittivity of free space, permittivity of vacuum
Dielektrizitätszahl relative permittivity
Diesel diesel
Diethylether [C] diethyl ether
Differenz [M] difference
Differenzial- und Integralrechnung [M] calculus
Differenzialrechnung [M] differential calculus
differenzieren [M] to derive
diffuse Reflexion diffuse reflection
Diffusor diffuser
Dimension dimension
Dimmer dimmer
Diode diode
Dioptrie dioptre
Dipol dipole
Dispersion dispersion
Dispersion (chromatische Dispersion von Licht) dispersion (chromatic dispersion of light)
divergent divergent
dividieren [M] to divide
Division [M] division
Doppelbindung [C] double bond
Doppelbrechung double refraction
Doppelspalt double slit
Doppelverglasung double glazing
doppelwandig double-walled
Dosenbarometer aneroid barometer
Drachen kite
Draht wire
Drahtseil steel cable, steel rope
Drehbewegung rotation
Drehkondensator variable capacitor
Drehmoment moment [rare], torque
Drehmomentschlüssel torque wrench
Drehsinn direction of rotation
Drehtisch rotary table
Drehung rotation
Drehzahlmesser rev counter, revolution counter, tachometer

dreiadriges Kabel three wire cable
Dreieck [M] triangle
Dreifingerregel three finger rule
Dreifingerregel der rechten Hand three finger rule of the right hand
Dreisatz [M] rule of three
dritter Hauptsatz der Thermodynamik third law of thermodynamics
drittes Kepler'sches Gesetz third Kepler law
drittes Newton'sches Gesetz third Newtonian law
Druck pressure
Druck ausüben to exert pressure
Druckerpatrone ink cartridge
Druckfeder compression spring
Druckluft compressed air
Druckverfahren printing method
Druckwelle pressure wave
Dubnium dubnium [Db, Z=105]
dunkel dark
Durchschnittsbeschleunigung average acceleration
Durchschnittsgeschwindigkeit average speed
Dur-Tonleiter major scale
Düse nozzle
Dynamik dynamics
dynamische Viskosität dynamic viscosity
Dynamo dynamo
dynamoelektrisches Prinzip dynamo-electric principle
Dysprosium dysprosium [Dy, Z=66]
Ebbe low tide
Echolot sonar
Effizienz efficiency
Eichenholz oak wood
Eigenfrequenz eigenfrequency, natural frequency
Eigenlänge proper length
Eigenschaft property
Eigenschwingung natural oscillation
Eigenzeit proper time
eindeutig unambiguous, unique
Einfachspalt single slit
Einfall (von Licht) incidence
einfallender Strahl incident ray
Einfallswinkel incidence angle
Einheit unit
Einheitskreis [M] unit circle
Einsteinium einsteinium [Es, Z=99]
Einstrahlung (Leistung pro Fläche) irradiance (power per area)
eintauchen to immerse
Eisberg iceberg
Eisen iron [Fe, Z=26]
Eisenbahnstrecke railway line
Eisenbahnwagen train carriage
Eisenbahnwagon train wagon
Eisenfeilspäne iron filings
Eisenkern magnetic core
Eisenspäne iron filings
Eisenträger iron girder
elastische Deformation elastic deformation
Elastizität elasticity
Elastizitätsmodul modulus of elasticity
elektrische Feldkonstante electric constant, electric field constant [rare], permittivity of free space, permittivity of vacuum
elektrische Feldlinie electric field line
elektrische Feldstärke electric field strength
elektrische Kraft electric force
elektrische Ladung electric charge
elektrische Leistung electric power
elektrische Stromstärke electric current
elektrischer Heizkörper electric radiator
elektrischer Schlag electric shock
elektrischer Strom electric current
elektrischer Stromkreis electric circuit
elektrischer Widerstand (Bauteil) electric resistor, electrical resistor
elektrischer Widerstand (physikalische Größe) electric resistance, electrical resistance
elektrisches Bauteil electrical component
elektrisches Feld electric field
Elektrizität electricity
Elektrode electrode
Elektrodynamik electrodynamics
elektromagnetische Kraft electromagnetic force
elektromagnetisches Spektrum electromagnetic spectrum
Elektromagnetismus electromagnetism
Elektromotor electric motor
Elektron electron
Elektronenanregung electron excitation
Elektronendichte electron density
Elektronenlawine electron avalanche
Elektronenvolt [Einh.] electron volt
Elektronenzahl electron number
elektronisches Signal electronic signal
Elektroskop electroscope
Elektrostatik electrostatics
Element [C] element
Elementarladung elementary charge
Elementarmagnet basic magnet
Elementname [C] element name
Elementsymbol [C] chemical symbol
Ellipse ellipse
elliptisch elliptic
Elongation deflection, elongation
Emission emission
Emissivität emissivity
emittieren to emit

Endpunkt eines Vektors [M] final point of a vector, tip of a vector
Energie energy
Energie konsumieren to consume energy
Energie verbrauchen to use energy
Energieausbeute energy gain, energy yield
Energiedichte energy density
Energiedosis absorbed dose
Energieerhaltungssatz energy conservation law
Energiefluss energy flow
Energiegewinn energy gain, energy yield
Energieniveau energy level
energiereich energy-rich
Energieträger carrier of energy, energy carrier
Energieübertragung energy transfer
Energieumwandlung energy transformation
Energieumwandlungskette energy transformation chain
Energieverlust energy loss
entgegengesetztes Vorzeichen opposite sign
entgleisen to derail
Epizentrum epicentre
Erbium erbium [Er, Z=68]
Erdbeben earthquake
Erde Earth
Erde (elektrisches Kabel) earth (electric wire) [B. E.], ground (electric wire) [A. E.], ground wire [A. E.]
erden to earth [B. E.], to ground [A. E.]
Erden earthing [B. E.], grounding [A. E.]
Erdgas [C] natural gas
Erdkabel earth wire [B. E.]
Erdkern Earth's core
Erdkruste Earth's crust
Erdmagnetfeld Earth's magnetic field
Erdmagnetismus geomagnetism
Erdmantel Earth's mantle
Erdung earthing [B. E.], ground [A. E.], grounding [A. E.]
erlaubte Bahn allowed orbit
Erlenmeyerkolben Erlenmeyer flask
Erreger (einer Schwingung) exciter (of an oscillation)
Erreger (einer Welle) exciter (of a wave)
Ersatzteil spare part
erstarren to freeze
erster Hauptsatz der Thermodynamik first law of thermodynamics
erstes Kepler'sches Gesetz first Kepler law
erstes Newton'sches Gesetz first Newtonian law
erwünschte Reibung desirable friction
Erz ore
erzwungene Schwingung forced oscillation

Esslöffel [Einh.] tablespoon
Ethan [C] ethane
Ethanol [C] ethanol
etwas aufwickeln to coil up something, to wind up something
Europäisches Verbundnetz European integrated network
Europium europium [Eu, Z=63]
eustachische Röhre (Ohr) Eustachian tube (ear)
Expansion expansion
Experiment experiment
Explosion explosion
Exponent [M] exponent
Exponentialfunktion [M] exponential function
externe Verbrennung external combustion
exzentrisch eccentric
Exzentrizität eccentricity
Fadenkreuz reticle
Fadenpendel simple pendulum, string pendulum
Fahrplan timetable
Fahrraddynamo bicycle dynamo
Fahrradpumpe bicycle pump
Faktor [M] factor
faktorisieren [M] to factorize
Fallbeschleunigung acceleration due to gravity, gravitational field strength
Fallbeschleunigung an der Oberfläche surface gravity
Fallbeschleunigung an der Oberfläche am Äquator equatorial surface gravity
Fallschirm parachute
Farad [Einh.] farad
Faraday-Käfig Faraday cage
Farbaddition colour addition
Farbe colour
zwei Farben ergeben eine dritte two colours form a third, two colours produce a third, two colours yield a third
Farbfilter colour filter
farbiges Licht coloured light
farbiges Licht mischen to mix coloured light
Farbsubtraktion colour subtraction
Fauna fauna
Feder spring
Federarbeit work performed on a spring
Federenergie spring energy
Federkonstante spring constant
Federkraft spring force
federn to bounce, to spring
Federpendel spring pendulum
Federpendelmechanismus spring pendulum mechanism
Federung suspension

Federwaage spring scale
Fehlerstromschutzschalter ALCI (appliance leakage current interrupter), appliance leakage current interrupter, GFCI (ground fault circuit interrupter), ground fault circuit interrupter, RCD (residual current device), residual current device
Feinunze [Einh.] troy ounce
Feldstärke field strength
Felge rim, wheel rim
Fensterglas window glass
Fensterscheibe window pane
Fermium fermium [Fm, Z=100]
ferner Weltraum outer space
fernes Weltall outer space
Fernglas binoculars
Fernrohr telescope
Ferromagnet ferromagnet
ferromagnetisch ferromagnetic
ferromagnetische Substanz ferromagnetic substance
Ferromagnetismus ferromagnetism
fest solid
festes Ende fixed end
Fett grease
fettig greasy
Fetttröpfchen fat droplet
Feuchtigkeit moisture
Feuerstein flint
Feuerwehrschlauch fire hose
Fichtenholz spruce wood
FI-Schutzschalter ALCI (appliance leakage current interrupter), appliance leakage current interrupter, GFCI (ground fault circuit interrupter), ground fault circuit interrupter, RCD (residual current device), residual current device
FI-Sicherung ALCI (appliance leakage current interrupter), appliance leakage current interrupter, GFCI (ground fault circuit interrupter), ground fault circuit interrupter, RCD (residual current device), residual current device
Fit fit
fitten to fit
Fitten fitting
flach [M] shallow
Fläche area
Flaschenzug hoist
Flintglas flint glass
Flora flora
Flugbahn path
Flugdauer flight time
Flugzeug aeroplane [B. E.], airplane [A. E.]
Fluid fluid
Fluor fluorine [F, Z=9]
Fluss flow
flüssig liquid
Flüssigkeit liquid
Flüssigkeiten liquids
Flüssigkeitsthermometer liquid thermometer
Flut high tide
Förderband conveyor belt
Form shape
Fortbewegung locomotion
Fortpflanzungsgeschwindigkeit propagation speed
fossil fossil
Fotoplatte photographic plate
Fourier-Analyse [M] Fourier analysis
Fracht cargo
Francium francium [Fr, Z=87]
Fraunhoferlinie Fraunhofer line
freier Fall free fall
freies Elektron free electron
freies Ende free end
Fremdleuchter non-luminous object
Freon [C] freon
Frequenz frequency
Frequenzanteil frequency component
fundamentale Konstante fundamental constant
Funktionsprinzip function principle
Funktionsweise functionality
Funktionswert [M] function value
Fuß [Einh.] foot
Gadolinium gadolinium [Gd, Z=64]
Galileisches Fernrohr Galilean telescope, Galileo's telescope
Galilei-Transformation Galilean transformation
Gallium gallium [Ga, Z=31]
Gallone [Einh.] gallon
Gammaradioaktivität gamma radioactivity
Gammateilchen gamma particle
Gammazerfall gamma decay
Gangunterschied path difference
ganze Zahl [M] integer
Gas gas
Gasflasche gas cylinder
gasförmig gaseous
Gasteilchen gas particle
GAU worst case scenario
Gay-Lussac'sches Gesetz Gay-Lussac's law
gebunden [C] combined
gebunden an das Gitter [C] combined with the lattice
gebundenes Elektron bound electron
gedämpfte Schwingung damped oscillation
geerdet earthed [B. E.], grounded [A. E.]
Gefrieren freezing
Gefrierpunkt freezing point
Gegenfeld counter field
Gegenkathete [M] opposite

Gegenkraft counterforce
gegenphasig opposite in phase
sich gegenseitig abstoßen to repel each other
sich gegenseitig anziehen to attract each other
sich gegenseitig auslöschen to cancel each other out
Gegenspannung counter-voltage
Gegenstand object
Gegenstandsgröße object size
Gegenstandsweite object distance
Gehäuse housing
gehobelt planed
Gehörgang ear canal
Gehörknöchelchen ear ossicles
Gehörschädigung hearing damage
Gehörschnecke cochlea
Geiger-Müller-Zähler Geiger-Müller tube
gekoppelte Oszillatoren coupled oscillators
gelb yellow
genau precise
Generator generator
Generatorspannung generator voltage
Generatorstrom generator current
Geomagnetismus geomagnetism
Geometrie [M] geometry
geometrisches Objekt [M] geometric object
geostationär geostationary
geothermisches Kraftwerk geothermal power station
Gerade [M] straight line
gerade Welle straight wave
geradeaus verlaufen to proceed straight
Gerät gadget
Germanium germanium [Ge, Z=32]
Germaniumdiode germanium diode
Gesamtdruck total pressure
Gesamtenergie total energy
geschlossener Stromkreis closed circuit, closed electric circuit
geschmolzen (z. B. Eisen) molten (e. g. iron)
Geschwindigkeit speed
Geschwindigkeitsamplitude speed amplitude
Geschwindigkeitsmesser speedometer
Geschwindigkeitsvektor velocity
Gesetz des Archimedes law of Archimedes
Gesetz von Pascal Pascal's law
a geteilt durch b [M] a divided by b, a over b
Gewicht weight
Gewinde thread
ein Gewinde schneiden to cut a thread
gewölbt arched, curved
Gezeiten tides
Gezeiten- tidal
Gitarre guitar
Gitarrensaite guitar string

Gitter diffraction grating, grating
Gitter (eines Festkörpers) lattice
glänzend shiny
Glas glass
Glasfaser fibreglass
Glasrohr glass tube
glatte Oberfläche smooth surface
Glatteis black ice
gleich sein [M] to equal
gleichförmige Bewegung uniform motion
gleichförmige Bewegung mit Anfangspunkt uniform motion with starting point
gleichförmige Kreisbewegung uniform circular motion
Gleichgewicht equilibrium
Gleichgewichtslage equilibrium position
gleichmäßig beschleunigte Bewegung uniformly accelerated motion
gleichmäßig beschleunigte Bewegung mit Anfangsgeschwindigkeit uniformly accelerated motion with initial speed
gleichmäßig beschleunigte Bewegung mit Anfangsgeschwindigkeit und Anfangspunkt uniformly accelerated motion with initial speed and with starting point
gleichnamige Pole like poles
Gleichspannung DC voltage, direct current voltage, direct voltage
Gleichstrom DC, DC current, direct current
Gleichstrommotor DC motor
Gleichung [M] equation
Gleichzeitigkeit concurrency
Gleisanlage railway track
gleiten to slide
Gleitkontakt sliding contact
Gleitreibung dynamic friction
Gleitreibungskraft dynamic friction force
Gleitreibungszahl dynamic friction coefficient
Glimmer mica
Glimmlampe glow lamp
glitschig greasy
Globales Förderband great ocean conveyor belt
Globus globe
Glühbirne incandescent bulb, incandescent light bulb, light bulb
eine Glühbirne leuchtet a light bulb shines
glühen to glow
Glühfaden (einer Glühbirne) filament (of a light bulb)
Glühkathode glow cathode
Glycerin glycerin, glycerine, glycerol
Glycerol glycerin, glycerine, glycerol
Glyzerin glycerin, glycerine, glycerol
Gold gold [Au, Z=79]
Golfstrom Gulf Stream

Deutsch - Englisch

Grad [M] degree
Grad Celsius [Einh.] degree Celsius
Grad Fahrenheit [Einh.] degree Fahrenheit
Gran [Einh.] grain
Granit granite
Graph [M] graph
graphisch darstellen [M] to graph, to plot
Graphit graphite
Grauguss cast iron
Gravitation gravitation
Gravitation (bezogen auf die Erde) gravity
Gravitationsfeldstärke gravitational field strength
Gravitationsgesetz law of gravitation
Gravitationskonstante gravitational constant
Gravitationskraft gravitational force
Gray [Einh.] gray
Grenzfläche boundary
Grieskorn semolina grain
große Bahnhalbachse semimajor axis
Größenordnung order of magnitude
größter annehmbarer Unfall worst case scenario
Grundfarbe primary colour
Grundlage basis
Grundschwingung fundamental, fundamental oscillation
Grundton fundamental, fundamental tone
Grundzustand ground level, ground state
Gummi rubber
Gummiband rubber band
Gusseisen cast steal
Güterwagen boxcar
Hafnium hafnium [Hf, Z=72]
haften to adhere
Haftreibung static friction
Haftreibungskraft static friction force
Haftreibungszahl static friction coefficient
Haftung adhesion
halbieren to halve
Halbkreis [M] semicircle
halbkreisförmig [M] semicircular
Halbleiter semiconductor
Halbring half ring
Halbschatten penumbra
Halbton semitone
Halbwertszeit half life
Hall-Effekt Hall effect
Hall-Sonde Hall probe
Halogenlampe halogen lamp
Halteposition stop position
Halter (z. B. für eine Spaltblende) holder (e. g. for a slit aperture)
Hammer (Ohr) hammer (ear)
Hammer (Werkzeug) hammer (tool)
Hammerkopf hammer head

Hammerstiel hammer handle
Handbremse handbrake
Handwerker craftsman
harmonische Schwingung harmonic oscillation, SHM (simple harmonic motion), simple harmonic motion
harmonische Welle harmonic wave
Hartgummi hard rubber
Hassium hassium [Hs, Z=108]
Hauptmaxima main maxima
Hauptmaximum main maximum
Hauptschalter main switch
Hauptzerfallsmodus main decay mode
Hebebühne hydraulic lift
Heisenberg'sche Unschärferelation Heisenberg uncertainty principle
Heizdraht heating wire
Heizkörper radiator
Heizwert calorific value, heat of combustion, heating value, lower heating value, net calorific value, net heat of combustion
Hektar [Einh.] hectare
Helium helium [He, Z=2]
hell bright
Helligkeit brightness
Henry [Einh.] henry
Heptan [C] heptane
Herd cooker [B. E.], stove
Herdplatte hotplate
herleiten to deduce
Herleitung deduction
Hertz [Einh.] hertz
Herzflimmern cardiac fibrillation
Herzschlag heartbeat
Hexan [C] hexane
Himmelskörper celestial object
hinterfragen to question, to scrutinise
x hoch n [M] x to the nth power, x to the power of n
Hochspannung high voltage
Hochspannungsleitung transmission line
Höhe height
Höhenmesser altimeter
Hohlspiegel concave mirror
Hohlspiegelgleichung concave mirror equation
Holmium holmium [Ho, Z=67]
Holz wood
Holzschraube screw
homogen homogeneous, homogenous
Hooke'sches Gesetz Hooke's law
Hörnerv cochlear nerve
Hornhaut (Auge) cornea (eye)
Hörschwelle hearing threshold
Hubarbeit lifting work
Hubraum cylinder capacity

Hufeisenmagnet horseshoe magnet
Hydraulik hydraulics
hydraulisch hydraulic
hydraulische Presse hydraulic press
Hydrostatik hydrostatics
hydrostatisches Paradoxon hydrostatic paradox
Hyperbel [M] hyperbola
Hypotenuse [M] hypotenuse
Hypothese hypothesis
Hypothesen hypotheses
Hypozentrum hypocentre
im Gegenuhrzeigersinn anti-clockwise [adj.] [adv.] [B. E.], counterclockwise [adj.] [adv.] [A. E.], in a counterclockwise direction [A. E.], in an anti-clockwise direction [B. E.]
im Uhrzeigersinn clockwise [adj.] [adv.], in a clockwise direction
Impuls momentum
in Parallel in parallel
in Phase in phase
in Reihe in series
in Reihe geschaltet connected in series
Index [M] index
Indium indium [In, Z=49]
Indizes [M] indices
Induktion electromagnetic induction, induction
Induktionsherd induction cooker [B. E.], induction stove
Induktionsspannung induction voltage
Induktivität inductance
induzierte Spannung induced voltage
Inertialsystem inertial system
Influenz electrostatic induction
infrarotes Licht infrared light
Infraschall infrasound
Infraschall- infrasonic
Inklination inclination
Inkompressibilität incompressibility
nach innen inwards [adj.] [adv.]
Innenohr inner ear
Innenwiderstand internal resistance
Innenwiderstand des menschlichen Körpers internal resistance of the human body
innere Energie internal energy
Inneres interior
instabil unstable
Integral [M] integral
Integralrechnung [M] integral calculus
Integration [M] integration
integrieren [M] to integrate
Intensität intensity
Interferenz interference
interferieren to interfere
internationaler Farbcode international colour code
internationaler Widerstandscode international colour code
interne Verbrennung internal combustion
Intervall [M] interval
Invar invar
inverse Galilei-Transformation inverse Galilean transformation
Iod iodine [I, Z=53]
Ion ion
Ionenantrieb ion propulsion
Ionenbindung [C] ionic bond
Ionisation ionisation
Iridium iridium [Ir, Z=77]
Iris iris
isobar isobaric
isochor isochoric
Isolation (elektrisch) insulation (electrical)
Isolation (thermisch) insulation (thermal)
Isolator (elektrisch) insulator (electrical)
Isolator (thermisch) insulator (thermal)
isolieren (elektrisch) to insulate (electrically)
isolieren (thermisch) to insulate (thermally)
isotherm isothermal
Isotop isotope
jemanden röntgen to take an X-ray of someone
Joule [Einh.] joule
Julianisches Jahr [Einh.] Julian year
Jupiter Jupiter
justieren to adjust
Kadmium cadmium [Cd, Z=48]
kalibrieren to calibrate
kalibriert calibrated
Kalium potassium [K, Z=19]
Kalk lime
Kalkstein limestone
Kalorie [Einh.] calorie
Kammerton chamber pitch
Kantenlänge [M] edge length
Kapazität (einer Batterie) capacity (of a battery)
Kapazität (eines Kondensators) capacitance
Kapsel capsule
Karat [Einh.] carat
Karbonstahl carbon steel
Kathode cathode
Kathodenstrahlrohr cathode ray tube
Kation [C] cation
Kehrwert [M] reciprocal
Keilriemen V-belt
Kelvin [Einh.] kelvin
Kennlinie characteristic, characteristic curve
Kepler'sches Gesetz Kepler law
Kepler'sches Fernrohr Kepler telescope
Keramik ceramics
Kern nucleus

Kerne nuclei
Kernfusion nuclear fusion
Kernkraft nuclear force
Kernkraftwerk nuclear plant, nuclear power plant, nuclear power station
Kernschatten umbra
Kernspaltung nuclear fission
Kerosin kerosene
Kerzenhalter candle holder
Kette chain
Kies gravel
Kilogramm [Einh.] kilogram
Kilogramm pro Kubikmeter [Einh.] kilogram per cubic metre
Kilometer pro Stunde [Einh.] kilometre per hour
Kilometerzähler mileometer, milometer [rare], odometer [A. E.]
Kilowattstunde [Einh.] kilowatt-hour
Kinematik kinematics
kinetische Energie KE (kinetic energy), kinetic energy
kinetische Energie eines Gasteilchens kinetic energy of a gas particle
Kirchhoffsche Gesetze Kirchhoff's circuit laws
Kirchhoffsche Knotenregel KCL (Kirchhoff's current law), Kirchhoff's current law
Kirchhoffsche Maschenregel Kirchhoff's voltage law, KVL (Kirchhoff's voltage law)
Klang timbre
Klangfarbe acoustic colour
klassische Relativität classical relativity
Klaviersaite piano string
Klotz block
Knete modelling clay, plasticine
Knetmasse plasticine
Knochen bone
Knoten [Einh.] knot
Knoten node
Knotenlinie nodal line
Knotenregel KCL (Kirchhoff's current law), Kirchhoff's current law
Kobalt cobalt [Co, Z=27]
Kochsalz [C] table salt
Kochsalzkristall [C] table salt crystal
Kohle carbon
Kohlekraftwerk coal-fired power station
Kohlendioxid [C] carbon dioxide
Kohlenmonoxid [C] carbon monoxide
Kohlenstoff carbon [C, Z=6]
Kolben piston
Kollimation collimation
Kollimator collimator
Komet comet
kommunizierende Röhren communicating vessels

Kommutator commutator
Kompass compass
Kompassnadel compass needle
Komplementärfarben complementary colours
komplexe Zahl [M] complex number
Kompressibilität compressibility
Kondensation condensation
Kondensationskern condensation nucleus
Kondensator capacitor
Kondensatorplatte capacitor plate
kondensieren to condense
Konkavlinse concave lens
Konstantan constantan
konstante Funktion [M] constant function
konstruktive Interferenz constructive interference
Kontaktkraft contact force
Konvektion convection
Konvektionsrohr convection tube
konventionelle Stromrichtung conventional current direction
konvergent convergent
Konvexlinse convex lens
Konzept concept
Koordinatensystem [M] coordinate system
Kopplung coupling
Kork cork
Körper body, object
Körperfarben body colours
kosmische Strahlung cosmic radiation
kovalente Bindung [C] covalent bond
Kraft force
eine Kraft ausüben to exert a force
Kräftegleichgewicht balance of forces, equilibrium of forces
Kraftstoß impulse
Kraftwerk power plant, power station
Kraftwerksgenerator power generator
Krampf cramp
Kran crane
Kreis [M] circle
Kreisbahn circular path
Kreisbeschleuniger accelerator ring
Kreisbewegung circular motion
Kreisdiagramm pie chart
Kreisfrequenz angular frequency
Kreisstrom circular current
Kreiswelle circular wave
Kreuzschraubendreher Phillips screwdriver
Kristall crystal
Kristallgitter crystal lattice, lattice
kritischer Winkel der Totalreflexion critical angle of total reflection
Krokodilklemme crocodile clamp
Krypton krypton [Kr, Z=36]
Kubikmeter [Einh.] cubic metre

Kubikmeter pro Sekunde [Einh.] cubic metre per second
Kuchendiagramm pie chart
Kugel ball, sphere
Kugel [M] sphere
Kugel (Geschoss) bullet
Kugellager ball bearing
Kugelwelle spheric wave
Kuiper-Gürtel Kuiper belt
Kundt'sches Rohr Kundt's tube
künstliche Quelle radioaktiver Strahlung artificial radiation source
Kupfer copper [Cu, Z=29]
Kuppel cupola
Kupplung clutch
Kupplungsscheibe clutch disk
Kurbel crank
kurbeln to crank
Kurbelwelle crankshaft
Kurve bend, curve
Kurve [M] graph
um eine Kurve fahren to go around a bend, to go round a curve
kurzschließen to short-circuit
Kurzschluss short, short circuit, short-circuit
kurzsichtig myopic, short-sighted
Kurzsichtigkeit short-sightedness
Lack lacquer
laden to charge
Ladung charge
Ladungs-Massenverhältnis des Elektrons charge-to-mass ratio for the electron
Ladungzahl charge number
Lagrangepunkt Lagrangian point
laminare Strömung laminar flow
Lampe lamp
Lampenaufhängung lamp mounting, light fixture
Landebahn runway
Länge length
Länge eines Vektors [M] length of a vector, magnitude of a vector, norm of a vector
Längenausdehnung linear thermal expansion
Längenausdehnungskoeffizient linear thermal expansion coefficient
Längendichte (einer Gitarrensaite) linear density (of a guitar string)
Längengrad longitude
Längenkontraktion length contraction
Langstreckenrakete long-range missile
Lanthan lanthanum [La, Z=57]
latente Wärme latent heat
Lautsprecher loudspeaker
Lautstärkepegel loudness
Lawrencium lawrencium [Lr, Z=103]
Leckstrom leakage current

LED (lichtemittierende Diode) LED (light emitting diode)
Leder leather
leere Batterie flat battery
Legende (eines Diagramms) legend (of a diagram)
Legierung alloy
Leinsamenöl linseed oil
Leistung power
Leistung einer Welle power of a wave
Leistung freisetzen to release power
Leistung in einem Widerstand power at a resistor
Leistung konsumieren to consume power
Leistung zur Verfügung stellen to provide power
Leistungsbeziehung beim Transformator transformer power relation
Leistungsverlust power loss
leiten (elektrisch) to conduct (electrically)
leiten (thermisch) to conduct (thermally)
Leiter (elektrisch) conductor (electrical)
Leiter (thermisch) conductor (thermal)
Leiterdraht conducting wire
Leiterschaukel conductor swing
Leiterschlaufe wire loop
Leitung (elektrisch) conduction (electrical)
Leitung (thermisch) conduction (thermal)
Lenkstange (Fahrrad) handlebars
Lenz'sche Regel Lenz's rule
Leuchtdiode LED (light emitting diode), light emitting diode
leuchten (z. B. eine Glühbirne) to shine (e. g. a light bulb)
Leuchtschicht (einer Leuchtstoffröhre) luminescent layer (of a fluorescent lamp)
Leuchtstoffröhre fluorescent lamp
Lichtausbreitung light propagation
Lichtgeschwindigkeit im Vakuum [Einh.] speed of light in vacuum
Lichtgeschwindigkeit in einem Medium speed of light in a medium
Lichtjahr [Einh.] light-year
Lichtmaschine (Auto) dynamo
Lichtquelle light source
Lichtschranke light barrier
Lichtstrahl light ray
Lichtstrahlen treffen aufeinander light rays coincide
Lichtstrahlen treffen sich light rays coincide
Lineal ruler
Linearbeschleuniger linear accelerator
lineare Bewegung linear motion
lineare Funktion [M] linear function
Linienspektren line spectra
Linienspektrum line spectrum

Linse

Linse lens
Linsenabstand lens distance
Linsengleichung lens equation
Linsenkombination lens combination
Linsensystem lens system
Liter [Einh.] litre
Lithium lithium [Li, Z=3]
Lochkamera pinhole camera
Lochscheibe punched disc
Lochsirene siren disc
Logarithmus [M] logarithm
Logarithmus von a zur Basis b [M] logarithm of a to base b
logisch [M] logical
Lokomotive locomotive
longitudinale Kopplung longitudinal coupling
longitudinale Welle longitudinal wave
Lorentz-Faktor Lorentz factor
Lorentzkraft Lorentz force
löten to solder
Lötkolben soldering iron
Luft air
luftbereift pneumatic-tyred
Luftdruck air pressure
Luftfeuchtigkeit humidity
Luftkissen air bearing, air cushion
Luftkissenschiff hovercraft
Luftkissentisch air bearing stage
Luftpumpe air pump, tyre pump
Luftwiderstand air resistance
Luftzirkulation air circulation
Luminosität luminous intensity
Lunge lung
Lupe magnifying glass
Lutetium lutetium [Lu, Z=71]
Magdeburger Kugel Magdeburg sphere
magenta magenta
Magma magma
Magnesium magnesium [Mg, Z=12]
Magnetfeld magnetic field
Magnetfeldlinie magnetic field line
Magnetfeldstärke magnetic field strength
magnetische Feldkonstante permeability of vacuum
magnetische Feldlinie magnetic field line
magnetische Feldstärke magnetic field strength
magnetische Flussdichte magnetic flux density
magnetische Influenz magnetic induction
magnetische Kraft magnetic force
magnetische Reaktion magnetic response
magnetischer Fluss magnetic flux
magnetisieren to magnetise
magnetisiert magnetised
Magnetisierung magnetisation
Magnetismus magnetism

Magnetit magnetite
Makrokosmos macrocosm
makroskopisch macroscopic
mal [M] times
a mal b [M] a times b
Malkasten paintbox
Mangan manganese [Mn, Z=25]
Manometer manometer
manövrieren to manoeuver
Marmor marble
Mars Mars
Maschenregel Kirchhoff's voltage law, KVL (Kirchhoff's voltage law)
Maßband measuring tape
Masse mass
Masse-Energie-Äquivalenz mass-energy equivalence
Massenspektrometer mass spectrometer
Massenzahl mass number
Mast mast
Materialeigenschaft material property, property of a material
Materialien materials
Materialprüfung material control
Mathematik mathematics
matt matt
Maxima maxima
maximale Höhe maximum height
maximaler Wirkungsgrad maximum efficiency
Maximum maximum
Mechanik mechanics
Meereshöhe sea level
Meerwasser sea water
Mehrfachspaltblende multi-slit aperture
Meile [Einh.] mile
Meitnerium meitnerium [Mt, Z=109]
Mendelevium mendelevium [Md, Z=101]
Menge [M] set
Merkur Mercury
Messdaten measured data
Messdiagramm chart, diagram, graph
messen to measure
Messgerät measurement device, measuring device, measuring instrument
Messgröße measurand
Messing brass
Messschieber caliper, calliper, gauge
Messtabelle chart, spreadsheet, table
Messung measurement
Messwerte measured data
Metallbindung [C] metallic bond
metallischer Leiter metallic conductor
Metallschraube bolt
Meter [Einh.] metre
Meter pro Quadratsekunde [Einh.] metre per second squared

Meter pro Sekunde [Einh.] metre per second
Methan [C] methane
Methanol [C] methanol
Methoden methods
Mikrokosmos microcosm
Mikrophon microphone
Mikroskop microscope
mikroskopisch microscopic
Millimeter Quecksilbersäule [Einh.] millimetre of mercury
Minima minima
minimale Sehweite closest focusing distance
minimale Sichtdistanz closest focusing distance
minimieren to minimise
Minimierung minimisation
Minimum minimum
Minuend [M] minuend
minus [M] minus
a minus b [M] a minus b
a minus b ist gleich c [M] a minus b equals c
Minuspol minus pole
Minute [Einh.] minute
Mischfarbe secondary colour
Mischtemperatur temperature of the mixture
im Mittel on average
mitteln to average
Mittelohr middle ear
Mittelwert average
mittlere Massenzahl [C] average mass number
mittlerer Radius mean radius
Modell model
Modellvorstellung model concept
moduliert modulated
Mol [Einh.] mole
Molybdän molybdenum [Mo, Z=42]
Momentanbeschleunigung instantaneous acceleration
Momentangeschwindigkeit instantaneous speed
Mond moon
Morgen [Einh.] acre
Motor engine, motor
Motorrad motorcycle
Multiplikation [M] multiplication
multiplizieren [M] to multiply
Musik music
Muster template
Mutter nut
Mutterkern parent nucleus
nach außen outwards [adj.] [adv.]
nach innen inwards [adj.] [adv.]
Nachleuchten phosphorescence
Nadellager needle bearing
Nadelloch pin hole, pinhole

Näherung approximation
Näherungsformel approximate formula
Natrium sodium [Na, Z=11]
Natriumchlorid [C] sodium chloride
Natriumdampflampe sodium lamp
Natur nature
Naturgesetz law of nature
natürliche Radioaktivität background radiation, natural background radiation
natürliche Zahl [M] natural number
Naturwissenschaft natural science
nautische Meile [Einh.] nautical mile
Nebelkammer cloud chamber
negativ geladen negatively charged
negativer Pol negative pole
negatives Vorzeichen [M] negative sign
Neigung (einer schiefen Ebene) inclination (of an inclined plane)
Neigungswinkel (einer schiefen Ebene) inclination (of an inclined plane), inclination angle (of an inclined plane)
Nenner [M] denominator
Neodym neodymium [Nd, Z=60]
Neon neon [Ne, Z=10]
Neonlampe neon lamp
Neptun Neptune
Neptunium neptunium [Np, Z=93]
Netz net
Netzhaut retina
neu justieren to readjust
neutral neutral
Neutron neutron
Neutronenstern neutron star
Neutronenzahl neutron number
Newton [Einh.] newton
Newton'sche Gesetze Newtonian laws
Newtonmeter [Einh.] newton metre
Newton'sches Gravitationsgesetz Newton's law of gravitation
Nicht-SI-Einheit non-SI-Unit
Nickel nickel [Ni, Z=28]
Niob niobium [Nb, Z=41]
Nobelium nobelium [No, Z=102]
Nocke cam
Nockenwelle camshaft
Nomenklatur nomenclature
Nonan [C] nonane
nördliches Polarlicht aurora borealis, northern lights
Nordlicht aurora borealis, northern lights
Nordpol (der Erde) North Pole
Nordpol (eines Magneten) north pole
Normale normal
Normalkraft normal force
Normbedingungen standard conditions
Normdruck standard pressure

Normtemperatur standard temperature
Normwert standard value
Normwert der Fallbeschleunigung standard acceleration due to gravity
Normzustand standard state
Note (c, d, e ...) note (C, D, E ...)
Nukleon nucleon
Nuklid nuclide
Nuklidtafel table of nuclides
Nulldurchgang zero-crossing
Nullleiter (das Kabel) neutral wire
Nullpunkt [M] origin, zero point
Nullpunktgerade [M] straight line through zero
Nullvektor [M] null vector
Nutzlast payload
Nylon nylon
oberer Heizwert [alt] gross calorific value, gross heat of combustion, higher heating value
Oberfläche [M] surface area
Oberflächenspannung surface tension
Oberflächentemperatur surface temperature
Oberschwingung harmonic, overtone
Oberton overtone
Objekt body, object
Objektiv objective
Octan [C] octane
Ofen oven
offener Stromkreis open electric circuit
offenes System open system
Ohm [Einh.] ohm
Ohm'sches Gesetz Ohm's law
Ohmmeter ohmmeter
ohmscher Widerstand (das elektrische Bauteil) ohmic resistor
ohmscher Widerstand (die Eigenschaft) ohmic resistance
Ohr ear
Ohrenschützer ear muffs
Ohrenstopfen ear plugs
Ohrenstöpsel ear plugs
Ohrtrompete (Ohr) Eustachian tube (ear)
Oktave octave
Okular eyepiece
Olivenöl olive oil
Oortsche Kometenwolke Oort cloud
Opal opal
Optik optics
optimieren to optimise
optische Achse optical axis, principal axis
optische Bank optical bench
optische Dichte (Brechzahl) refractive index
optische Dichte (Extinktion) optical density
optische Dicke optical depth, optical thickness
optische Tiefe optical depth, optical thickness
optisches Instrument optical instrument

Ordnung des Hauptmaximums order of the main maximum
Ordnung des Maximums order of the maximum
Ordnungszahl [C] atomic number
Originalteil genuine part
O-Ring O-ring
Ort position
Ortsvektor position vector
Osmium osmium [Os, Z=76]
Oszillatorengeschwindigkeit oscillator velocity
ovales Fenster (Ohr) oval window (ear)
oxidieren to oxidise
oxidiert oxidised
Palladium palladium [Pd, Z=46]
Parabel [M] parabola
Parabolspiegel parabolic dish, parabolic mirror
Paraffin paraffin
Paraffinöl [C] paraffin oil
in Parallel in parallel
parallel parallel [adj.] [adv.]
parallel geschaltet connected in parallel
parallel zur Faser along the grain
Parallelogramm [M] parallelogram
Parallelschaltung parallel circuit, parallel connection
Parallelstrahl parallel ray
Paramagnet paramagnet
paramagnetisch paramagnetic
paramagnetische Substanz paramagnetic substance
Paramagnetismus paramagnetism
partielle Sonnenfinsternis partial solar eclipse
Pascal [Einh.] pascal
Pascalsekunde [Einh.] pascal second
Paukenhöhle (Ohr) middle ear cavity
Pechblende pitchblende
Pendel pendulum
Pendelkörper bob, pendulum bob
Pendeluhr pendulum clock
Pentan [C] pentane
Periodensystem [C] periodic system of the elements
periodisch periodic
Permanentmagnet permanent magnet
Permanentmagnetismus permanent magnetism
Permeabilitätszahl relative permeability
Permittivität des Vakuums [selten] electric constant, electric field constant [rare], permittivity of free space, permittivity of vacuum
Perpetuum mobile perpetuum mobile, perpetuum motion machine
Petroleum petroleum
Pfaddifferenz path difference

Pfeifenorgel pipe organ
Pfeil arrow
Pfeil und Bogen bow and arrow
Pfeilblende arrow aperture
Pferdestärke [Einh.] horsepower
Pfund [Einh.] pound
Phänomen phenomenon
Phänomene phenomena
Phase phase
Phase (das elektrische Kabel) phase (the phase wire), phase wire
Phase (einer Schwingung) phase (of an oscillation)
Phase (einer Welle) phase (of a wave)
phasenabhängige Amplitude phase-dependent amplitude
Phasendifferenz phase difference
Phasengeschwindigkeit celerity, phase velocity
phasenverschoben phase-shifted
Phosphor phosphorus [P, Z=15]
Phosphoreszenz phosphorescence
Photon photon
Photonenenergie photon energy
physikalische Formel physics formula
physikalische Konstante physical constant
physikalische Stromrichtung real current direction
Pi-Bindung [C] pi bond
Pint [Einh.] pint
Planck'sches Wirkungsquantum Planck constant
Planet planet
Planetenbahn planetary orbit
Planetenbewegung planetary motion
Planetoid planetoid
Planzeichnung plan
Plastik plastic
plastische Deformation plastic deformation
Platin platinum [Pt, Z=78]
Plattenabstand plate distance
Plattenfläche plate area
Plattenkondensator parallel plate capacitor
platter Reifen flat tyre
Pleuel piston rod
Pleuelstange piston rod
Plexiglas Plexiglas, plexiglass
plus [M] plus
a plus b [M] a plus b
Pluspol plus pole
Plutonium plutonium [Pu, Z=94]
pneumatisch pneumatic
Pol pole
polar polar
Polarisation polarisation
Polarität polarity
Polarlicht aurora

polieren to polish
poliert polished
Polonium polonium [Po, Z=84]
Polynom [M] polynomial
Porzellan porcelain
positiv geladen positively charged
positiver Pol positive pole
positives Vorzeichen [M] positive sign
Postulate der Relativitätstheorie postulates of relativity
Potenz [M] power
n-te Potenz von x [M] nth power of x
potenzielle Energie PE (potential energy), potential energy
Potenzieren [M] exponentiation
Potenziometer potentiometer
Pound-force (US-amerikanische Krafteinheit) [Einh.] pound-force
Präfix prefix
Präfixe prefixes
Präparat compound
Praseodym praseodymium [Pr, Z=59]
Presslufthammer pneumatic hammer
primäres Elektron primary electron
Primärkreis primary circuit
Primärspule primary coil
Primärteilchen primary particle
Primärwelle primary wave
Prinzip der Umkehrbarkeit des Lichtwegs principle of the reversibility of the light path
Prisma prism
Probeladung test charge
Produkt [M] product, quotient
Projektion projection
Promethium promethium [Pm, Z=61]
Propan [C] propane
proportional proportional
proportional zu proportional to
Proportionalität proportionality
Proportionalitätskonstante constant of proportionality, proportionality constant
Protactinium protactinium [Pa, Z=91]
Proton proton
Protonenzahl proton number
psi (US-amerikanische Druckeinheit) [Einh.] pound-force per square inch, psi
Puffer buffer
Pumpe pump
Punkt der maximalen Höhe point of maximum height
punktförmig point-like
punktförmige Quelle point source
Punktladung point charge
Quader [M] cuboid
Quadrat [M] square
Quadratmeter [Einh.] square metre

Quadratwurzel [M] square root
Quant quantum
Quantenmechanik quantum mechanics
Quantenphysik quantum physics
Quantensprung quantum leap
Quart [Einh.] quart
Quarz quartz
Quarzglas quartz glass
Quarzuhr quartz watch
Quecksilber mercury [Hg, Z=80]
Quecksilbersäule column of mercury
Quecksilberthermometer mercury thermometer
quer zur Faser against the grain
Querschnittsfläche cross sectional area
Quotient [M] quotient
Räder drehen durch wheels spin
radial radial
radial nach außen radially outwards [adv.]
radial nach innen radially inwards [adv.]
Radian [M] radian
Radikand [M] radicand
radioaktiv radioactive
radioaktive Strahlung nuclear radiation, radiation
radioaktiver Zerfall decomposition, disintegration, radioactive decay
radioaktives Isotop radioactive isotope
radioaktives Zerfallsgesetz radioactive decay law
Radioaktivität radioactivity
Radiokarbonmethode radiocarbon dating
Radium radium [Ra, Z=88]
Radius [M] radius
Radius der volumengleichen Kugel volumetric mean radius
radizieren [M] to extract a root
Radon radon [Rn, Z=86]
Rakete (Raumfahrt) rocket
Rakete (Waffe) missile
Raketenstufe rocket stage
Rapssamenöl rape seed oil
rationale Zahl [M] rational number
Rauchmelder smoke detector
raue Oberfläche rough surface
Rauheit roughness
Raumfahrt space flight
räumlich [M] spacial
Raumschiff spacecraft, spaceship
Raumwinkel [M] solid angle
Raupe caterpillar
Reagenzglas test tube
Reaktionsgleichung equation
Reaktionszeit reaction time
Rechteck [M] rectangle
Rechte-Hand-Regel right hand rule

rechtwinkliges Dreieck [M] right triangle [A. E.], right-angled triangle [B. E.]
reelle Funktion [M] real function
reelle Zahl [M] real number
reelles Bild real image
Reflektanz reflectance, reflectivity
reflektieren to reflect
reflektierter Strahl reflected ray
Reflexion reflection
Reflexionsgesetz law of reflection
Reflexionsgrad reflectance
Reflexionswinkel reflection angle
regelmäßige Reflexion regular reflection, specular reflection
Regeln rules
Regressionsgerade regression line
reiben to rub
Reibung friction
Reibungsarbeit friction work
reibungsfrei frictionless
Reibungszahl friction coefficient
Reifen tyre
Reifen blockieren wheels lock up
in Reihe in series
Reihenschaltung series circuit, series connection
Reiter (auf einer optischen Bank) slider (on an optical bench)
relative Längenänderung strain, tensile strain
relativistische Energie-Impuls-Beziehung relativistic energy-momentum relation
relativistische kinetische Energie relativistic kinetic energy
relativistische Masse relativistic mass
relativistischer Impuls relativistic momentum
Relativität relativity
Relativitätsprinzip principle of relativity, relativity principle
Relativitätstheorie theory of relativity
Resonanz resonance
Resonanzamplitude resonant value of the amplitude
Resonanzfrequenz resonant frequency
Resonanzkatastrophe resonance disaster
Resublimation deposition
resublimieren to deposit
resultierende Kraft resultant force
resultierende Schwingung resulting oscillation
Retina retina
Rhenium rhenium [Re, Z=75]
Rhodium rhodium [Rh, Z=45]
Richtung direction
Riemen belt
Riemenscheibe pulley
Ringbeschleuniger accelerator ring
Rizinusöl castor oil

Roboter robot
Rohöl crude oil
Rohr pipe, tube
Rolle pulley
Rollreibung rolling friction
Rollreibungskraft rolling friction force
Rollreibungszahl rolling friction coefficient
röntgen to X-ray
sich röntgen lassen to have an X-ray
Röntgenphoton X-ray photon
Röntgenquant X-ray quantum
Röntgenstrahl X-ray
Rotation rotation
Rotationsrichtung direction of rotation
Rotor armature, rotor
Rotverschiebung redshift
Rubidium rubidium [Rb, Z=37]
Rückholfeder return spring
Rückstoß recoil
rücktreibende Kraft restoring force
Ruder oar
Ruheenergie rest energy
Ruhelage rest position
Ruhemasse rest mass
ruhen to rest
Ruheposition stop position
Ruhesystem rest frame
Ruß soot
Ruthenium ruthenium [Ru, Z=44]
Rutherfordium rutherfordium [Rf, Z=104]
Säge saw
Sägemehl sawdust
sägen to saw
Saite string
Saiteninstrument string instrument
Salz [C] salt
Salzkristall [C] salt crystal
Salzwasser salt water, sea water
Samarium samarium [Sm, Z=62]
Sammellinse converging lens
Sand sand
Saphir sapphire
Satellit satellite
Satellitenschüssel satellite dish
Sattel (Gitarre) nut (guitar)
Saturn Saturn
Sauerstoff oxygen [O, Z=8]
Scandium scandium [Sc, Z=21]
Schall sound
Schall- sonic
Schalldruck sound pressure
Schalldruckpegel sound pressure level, SPL (sound pressure level)
Schallerzeuger sound generator
Schallpegel sound level, sound pressure level, SPL (sound pressure level)
Schallsender sonic sender
Schallwelle sound wave
Schaltbild circuit diagram
Schalter switch
Schaltplan circuit diagram
Schaltsymbol circuit symbol
scharf sharp
Schärfe sharpness
Schatten shadow
einen Schatten werfen to cast a shadow
Schattenbildungsgleichung shadow formation equation
Schattenraum shadow space
Scheibenbremse disk brake
Scheinkraft pseudo force
Scheinwerfer (Auto) head light (car)
Schere scissors
Scherkraft shear force
Scherwelle shear wave
Schiebewiderstand variable resistor
Schieblehre caliper, calliper, gauge
schiefe Ebene inclined plane
schiefer Wurf oblique throw [A. E.], projectile motion
Schiene rail, railway track
Schirm screen
Schlauch pipe, tube
Schlauch (groß) hose
Schlauchdruck hose pressure
Schleife loop
schleifen to grind
Schleifstein grinding stone, grindstone
Schleppschiff tug, tugboat
Schleudersitz catapult seat, ejection seat
Schlittschuh ice skate
schlussfolgern to conclude, to infer
schmelzen to melt
Schmelzen melting
Schmelzpunkt melting point
Schmerzgrenze pain limit
schmieren to lubricate
Schmiermittel lubricant
Schmirgelpapier sandpaper
Schnelle (Oszillatorengeschwindigkeit) oscillator velocity
Schnittfläche [M] intersection
Schnittpunkt [M] intersection
Schotter gravel
schräger Wurf oblique throw [A. E.], projectile motion
Schraubenbahn helical path
Schraubendreher screwdriver
Schraubenfeder coil spring
schraubenförmig helical
Schraubenlinie helix
Schraubenschlüssel wrench

Schraubstock vice
Schrödingergleichung Schrödinger equation
Schublehre caliper, calliper, gauge
schwach gedämpfte Schwingung lightly damped oscillation
schwache Dämpfung light damping
schwache Kernkraft weak nuclear force
schwache Kraft weak force
schwache Wechselwirkung weak interaction
schwarzes Loch black hole
schweben to levitate
schweben (in der Luft) to float
schweben (unter Wasser) to float
schweben lassen to levitate
Schwebung beat
Schwebungsfrequenz beat frequency
Schwefel sulphur [S, Z=16]
Schwefeldioxid [C] sulfur dioxide
Schwefelsäure [C] sulfuric acid
schweißen to weld
Schweißgerät welder
Schweredruck hydrostatic pressure
Schwerelosigkeit microgravity, weightlessness, zero gravity
schweres Wasser heavy water
Schwerkraft gravitation, gravitational force
Schwerkraft (bezogen auf die Erde) gravity
schwimmen (Lebewesen auf dem Wasser durch eigenen Krafteinsatz) to swim
schwimmen (Objekte oder Lebewesen, die auf dem Wasser treiben) to float
Schwingung oscillation
Schwingungsdauer period
schwingungsfähiges System vibratory system
Schwingungsgenerator vibration generator
Schwungrad flywheel
Seaborgium seaborgium [Sg, Z=106]
Segelflugzeug glider, sailplane
Sehwinkel field of view
Seil rope
Seilwinde winch
seismische Welle seismic wave
Seismogramm seismogram
sekundärer Regenbogen secondary rainbow
sekundäres Elektron secondary electron
Sekundärkreis secondary circuit
Sekundärspule secondary coil
Sekundärteilchen secondary particle
Sekundärwelle secondary wave
Sekunde [Einh.] second
Selbstleuchter luminous object
Selen selenium [Se, Z=34]
senkrecht [adj.] perpendicular, vertical
senkrecht [adv.] perpendicular, perpendicularly, vertically
seriell geschaltet connected in series

SI-Basisgröße SI-base quantity
Sicherung fuse, safety fuse
eine Sicherung fliegt heraus a fuse blows
eine Sicherung ist herausgeflogen a fuse is blown
sichtbar machen (Feldlinien) to visualise (field lines)
sichtbares Licht visible light
Sichtweite visibility
Siedepunkt boiling point
SI-Einheit SI-unit
Sievert [Einh.] sievert
Sigma-Bindung [C] sigma bond
signifikante Stellen significant figures
SI-Grundeinheit SI-base unit
Silber silver [Ag, Z=47]
Silikonöl silicone oil
Silizium silicon [Si, Z=14]
Sinneszellen sensory cells
Sinus [M] sine, sinus
sinusförmig [M] sinusoidal
Sinusfunktion [M] sinusoid
Sinuskurve [M] sine curve, sinusoid
Sirene siren
Skala scale
Skalar scalar
Skalarmultiplikation [M] scalar multiplication
skalieren (einen Vektor) [M] to re-scale (a vector)
Skispringer ski jumper
Skizze sketch
Sockel (eines Stativs) base (of a stand)
Sojabohnenöl soy been oil, soya been oil
Solarkraftwerk solar power station
Sollwert nominal value
Sonne sun
Sonnenabstand (eines Planeten) solar distance (of a planet)
Sonnenwind solar wind
spaltbar fissile
Spaltblende slit aperture
Spannkraft tension
Spannung stress, tensile stress, voltage
Spannung an einen Widerstand anlegen to apply voltage to a resistor
Spannung anlegen to apply voltage
Spannung fällt an einem Widerstand ab voltage drops at a resistor
Spannungen bei der Parallelschaltung voltages in the parallel circuit
Spannungen bei der Reihenschaltung voltages in the series circuit
Spannungsabfall voltage drop
Spannungsbeziehung beim Transformator transformer voltage relation

Spannungsquelle voltage source
Spannungsteiler potential divider, voltage divider
Sparlampe energy-saving bulb
Spektralfarben spectral colours
Spektren spectra
Spektroskop spectroscope
Spektrum spectrum
Spektrum des sichtbaren Lichts spectrum of visible light
Spezialfall special case
spezielle Relativitätstheorie special theory of relativity
spezifische Schmelzwärme specific latent heat of fusion
spezifische Verdampfungswärme specific latent heat of vaporization
spezifische Wärme specific heat
spezifische Wärmekapazität specific heat capacity
spezifischer Widerstand resistivity, specific resistance
Spiegel mirror
Spiegelung reflection
Spirale spiral
spitz pointed, sharp, spiky
Spitze eines Vektors [M] tip of a vector
Splitt grit
sprengen to blast, to blow, to blow up, to bust
Spritze syringe
spröde brittle
Spule coil
Spule (sehr groß) solenoid
Spulenlänge coil length, length of a coil
Stab bar, pole, rod
Stäbchen (Auge) rods (eye)
Stabmagnet bar magnet
Stahl steel
Stahlseil steel cable, steel rope
Stange bar, pole, rod
starke Dämpfung heavy damping
starke Kernkraft strong nuclear force
starke Kraft strong force
starke Wechselwirkung strong interaction
Startbahn runway
starten to launch
Stativ stand
Stator stator
Staubsauger carpet cleaner [A. E.], Hoover [B. E.], vacuum cleaner
Staudamm dam
Staudruck dynamic pressure
Steckdose outlet, power outlet, power point, socket
Stecker plug
Stefan-Boltzmann-Gesetz Stefan-Boltzmann law
Stefan-Boltzmann-Konstante Stefan-Boltzmann constant
Steg (Gitarre) bridge (guitar)
stehende Welle standing wave, stationary wave
steif rigid
Steigbügel stirrup
Steigbügel (Ohr) stirrup (ear)
Steigung [M] gradient, slope
Steigung (einer schiefen Ebene) gradient (of an inclined plane), slope (of an inclined plane)
Steigungswinkel (einer schiefen Ebene) slope angle (of an inclined plane)
Steigzeit rising time
steil [M] steep
Stellschraube (zur Höheneinstellung) levelling screw
Steuerruder helm
Stickstoff nitrogen [N, Z=7]
Stiftloch pin hole, pinhole
Stimmband vocal cord
Stimmgabel tuning fork
Stoffmenge amount of substance
Stokes-Gesetz Stokes' law
Stopfen plug
Stoppuhr stop watch
Störung disturbance
Stoßdämpfer shock absorber
Stoßstange bumper
Strahldichte (Leistung pro Fläche pro Raumwinkel) radiance (power per area per solid angle)
Strahlung radiation
Strahlungsdurchgang radiative transfer
Strahlungsgürtel radiation belt
strecken (einen Vektor) [M] to stretch a vector
Streichinstrument string instrument, stringed instrument
Streuung dispersion, scattering
Stricknadel knitting needle
Stroboskop stroboscope
Strohhalm straw
Strom current
Stromlinie streamline
Stromnetz grid [B. E.], mains
Stromschlag electric shock
Strom-Spannungs-Kennlinie current-voltage characteristic, current-voltage characteristic curve
Stromstärkebeziehung beim Transformator transformer current relation
Stromteiler current divider
Strömungswiderstand drag, fluid resistance
Strontium strontium [Sr, Z=38]

strukturiert structured
Stunde [Einh.] hour
Styropor polystyrene, styrofoam
Sublimation sublimation
sublimieren to sublimate
Substanz substance
Subtrahend [M] subtrahend
subtrahieren [M] to subtract
Subtraktion [M] subtraction
subtraktive Mischung subtractive mixture
südliches Polarlicht aurora australis, southern lights
Südlicht aurora australis, southern lights
Südpol (der Erde) South Pole
Südpol (eines Magneten) south pole
Summand [M] addend
Summe [M] sum
Supernova supernova
Superposition superposition
Supraleiter superconductor
Süßwasser fresh water
Symbol symbol
Synchronmotor synchronous motor
System system
tabellieren to tabulate
tabelliert tabulated
Tag [Einh.] day
Takt (Motor) stroke
Tangens [M] tangent
tangential [adj.] [M] tangential
tangential [adv.] [M] tangentially
Tankuhr fuel gauge, petrol gauge
Tantal tantalum [Ta, Z=73]
tarieren to tare
Taschenlampe torch
Tauchsieder immersion heater
Technetium technetium [Tc, Z=43]
Technik technique
technische Anwendung technical application
technische Stromrichtung conventional current direction
Teelöffel [Einh.] teaspoon
Teer tar
teeren to tar
Teflon Teflon
Teil part
n-ter Teil [M] nth part
Teilchen particle
Teilchenbeschleuniger particle accelerator
Teilchenbeschleunigungsring particle accelerator ring
Teilchenmodell particle model
Teilchenspur particle track
Teilchenzahl particle number
Teilchenzahldichte particle number density
Teilmenge von [M] subset of

Tellur tellurium [Te, Z=52]
Temperatur temperature
Temperaturfühler temperature sensor
Terbium terbium [Tb, Z=65]
Tesla [Einh.] tesla
Tetrachlorkohlenstoff [C] carbon tetrachloride
Textilien textiles
Thallium thallium [Tl, Z=81]
Theorie theory
thermische Elektronen thermal electrons
Thermochromlack thermochrome lacquer
Thermochrompapier thermochrome paper
Thermodynamik thermodynamics
thermodynamische Daten thermodynamic data
thermodynamisches Gleichgewicht thermodynamic equilibrium
Thermographie thermography
Thermometer thermometer
Thermosflasche thermos flask
Thorium thorium [Th, Z=90]
Thulium thulium [Tm, Z=69]
Tiefe depth
Tintenpatrone (Drucker) ink cartridge
Tischtennisball ping-pong ball
Titan titanium [Ti, Z=22]
Tochterkern daughter nucleus
Todesgefahr danger of death
Toluol [C] toluol
Ton clay, tone
Tonerkartusche toner cartridge
Tonerpatrone toner cartridge
Tonerzeuger sound generator
Tonhöhe pitch, tone pitch
Tonleiter musical scale
Tonne [Einh.] metric tonne
totale Sonnenfinsternis full solar eclipse
Totalreflexion total internal reflection, total reflection
Trägermedium carrier medium
Tragfläche (Flugzeug) wing (aeroplane)
Trägheit inertia
Trajektorie trajectory
Tram tram
Trampolin trampoline
Transformator transformer
Transformatorengleichung transformer equation
Transformatorenstation transformer station
Transmissionsgrad transmittance
transversale Kopplung transverse coupling
transversale Welle transverse wave
treiben (Objekte oder Lebewesen auf dem Wasser) to float
Treibstoff fuel

Trigonometrie [M] trigonometry
Trockeneis dry ice
Trommelbremse drum brake
Trommelfell (Ohr) eardrum
turbulente Strömung turbulent flow
Turbulenz turbulence
türkis turquoise
Überdruck overpressure
Übergang transition
Übergangswiderstand der Haut skin resistance
Überlagerung superposition
übermitteln to transmit
Überrest (eines Sterns) remnant (of a star)
übersättigt supersaturated
übersättigter Dampf supersaturated vapour
übersättigter Wasserdampf supersaturated water vapour
übertragen to transfer
Uhr clock
Ultraschall ultrasound
Ultraschall- ultrasonic
ultraviolettes Licht ultraviolet light
um die eigene Achse rotieren to spin
um eine Kurve fahren to go around a bend, to go round a curve
U-Manometer U-tube manometer
Umfang circumference (especially of a circle)
Umfang [M] perimeter
umgekehrt proportional inversely proportional, reciprocally proportional
umgekehrt proportional zu inversely proportional to, reciprocally proportional to
umgekehrtes Bild inverted image
Umkehrbarkeit des Lichtwegs reversibility of the light path
umkreisen (einen Planeten) to orbit (a planet)
Umlaufbahn orbit
Umlaufdauer orbit period, time period
Umlenkrolle pulley
Umschalter toggle
unerwünschte Reibung undesirable friction
ungebunden [C] uncombined
ungedämpfte Schwingung undamped oscillation
ungleichnamige Pole unlike poles
Universalgenerator universal generator
Universalmotor universal motor
Universum universe
unkalibriert uncalibrated
unregelmäßige Reflexion irregular reflection
Unruh (einer Uhr) balance wheel
unscharf blurred
Unterdruck underpressure [rare], vacuum

unterer Heizwert [alt] lower heating value, net calorific value, net heat of combustion
untergehen (z. B. im Wasser) to sink (e. g. in water)
Untergrund substrate, substratum
Unterlage substrate, substratum
untersuchen to investigate
Unze [Einh.] ounce
Uran uranium [U, Z=92]
Uran-Blei-Methode uranium-lead method
Uranerz uranium ore
Urangestein uranium rock
Uranus Uranus
Urkilogramm prototype kilogram
Urmeter prototype metre bar
U-Rohr U-tube
Ursache, Vermittlung, Wirkung cause, transmission, effect
Vakuum vacuum
Vakuumröhre vacuum tube
Van de Graaff Generator van de Graaff generator
Vanadium vanadium [V, Z=23]
Van-Allen-Gürtel Van Allen belt
Variable variable
Vektor [M] vector
Vektor der Coulombkraft Coulomb force vector
Vektor der Durchschnittsbeschleunigung average acceleration vector
Vektor der Durchschnittsgeschwindigkeit average velocity
Vektor der Gravitationskraft gravitational force vector
Vektor der Momentangeschwindigkeit instantaneous velocity
Vektor der Normalkraft normal force vector
Vektor der Phasengeschwindigkeit phase velocity
Vektor der Winkelgeschwindigkeit angular velocity
einen Vektor stauchen [M] to shorten a vector
einen Vektor zerlegen [M] to decompose a vector
Vektoraddition [M] vector addition
Vektoranalysis [M] vector analysis
Vektorzerlegung [M] resolution of a vector, vector decomposition
Ventil valve
Venus Venus
Verallgemeinerung generalisation
verankern to anchor
sich verbinden mit [C] to combine with
Verbrennung combustion

Verdampfen vaporisation, to vaporise
Verdichten (Takt 2 eines Viertaktmotors) compression (stroke 2 of a four-stroke engine)
Verdichtungsstörung compression disturbance
verdoppeln to double
verdrängen to displace
verdrängt displaced
verdrängte Flüssigkeit displaced liquid
Verdünnungsstörung rarefaction disturbance
verformen to deform
Verformung deformation
Vergaser carburetor [A. E.], carburettor [B. E.]
vergrößertes Bild enlarged image
Vergrößerung magnification
verkalken to fur up
Verlauf (z. B. einer Spannung) course (e. g. of a voltage)
Vermutung presumption
vernachlässigbar negligible
vernachlässigen to neglect
verschieben [M] to displace, to shift
Verschiebungsvektor [M] displacement
Verschleißteil wearing part
verschlüsselt encrypted
verschoben [M] displaced, shifted
verschwimmen to blur
verschwommen blurred
verspiegelt metallized, mirrored
Versuchsaufbau experimental setup, experimental set-up
vertikaler Wurf vertical throw
sich verzahnen to interlock
verzahnen to interlock
Verzögerung deceleration
Vidi-Dose (eines Dosenbarometers) aneroid capsule (of an aneroid barometer), aneroid cell (of an aneroid barometer)
Viertaktmotor four-stroke engine
virtueller Brennpunkt virtual focal point
virtuelles Bild virtual image
viskoser Dämpfungsfaktor viscous damping coefficient
Viskosität viscosity
Vollkreis [M] full circle
Volt [Einh.] volt
Voltmeter voltmeter
Volumen volume
Volumenausdehnung volumetric thermal expansion
Volumenausdehnungskoeffizient volumetric thermal expansion coefficient
Vorgang process
Vorsilbe prefix
Vorwiderstand pre-resistor
Vorzeichen [M] sign
Waage balance, scale

waagrechte Ebene level plane
waagrechter Wurf horizontal throw
Wachskugel wax ball
Walze roller
Wärme heat
Wärmeausbreitung heat propagation
Wärmeausdehnung thermal expansion
Wärmeausdehnungskoeffizient thermal expansion coefficient
Wärmebildkamera thermal imaging camera
Wärmebrücke heat bridge
Wärmedurchgang heat transfer
Wärmedurchgangskoeffizient heat transfer coefficient
Wärmekapazität heat capacity
Wärmekraftmaschine heat engine
Wärmelehre thermodynamics
Wärmeleitfähigkeitskoeffizient heat conduction coefficient
Wärmeleitung heat conduction
Wärmepumpe heat pump
Wärmestrahlung thermal radiation
Wärmeströmung convection
Wasser verdrängen to displace water
Wasserdampf water vapour
Wasserhahn tap
Wasserkessel kettle
Wasserkraftwerk hydroelectric power station
Wasserreservoir water reservoir
Wasserstoff hydrogen [H, Z=1]
Wasserstofflampe hydrogen lamp
Wasserstrahl water jet
Wasserverdrängung water displacement
Wasserwaage level
Wasserwelle water wave
Watt [Einh.] watt
Watte cotton wool
Weber [Einh.] weber
Wechselspannung AC voltage, alternating voltage
Wechselstrom AC, AC current, alternating current
Wechselstrommotor AC motor
Wehneltröhre Wehnelt tube
weißer Zwerg white dwarf
weitsichtig hyperopic, long-sighted
Weitsichtigkeit long-sightedness
Welle wave
Wellenarten types of waves
Wellenbauch antinode
Wellenberg crest, wave crest
Wellenfront wave front
Wellengleichung wave equation
Wellenknoten node
Wellenlänge wavelength
Wellental trough, wave trough

Wellenwanne ripple tank
Wellenzahl wave number, wavenumber
Weltall space
Weltbild world view
Weltraum space
Wert value
Wickelkondensator wound capacitor
Widerstand (Bauteil) resistor
Widerstand (physikalische Größe) resistance
Widerstandsbeiwert drag coefficient
Widerstandscode colour code
Widerstandskombination resistor combination
Widerstandsthermometer resistance thermometer
Winde winch
Windkraftwerk wind power station, wind-driven power station
Windschatten slipstream
Windung (einer Spule) loop (of a coil)
Windungszahl number of loops
Winkel [M] angle
Winkel der Totalreflexion critical angle of total reflection
einen Winkel einschließen mit [M] to enclose an angle with, to encompass an angle with
Winkelgeschwindigkeit angular speed
Wirbel (Gitarre) tuning key (guitar)
Wirbelstrom eddy current
Wirbelstrombremse eddy current brake
Wirkungsgrad efficiency
Wissenschaft science
wissenschaftliche Schreibweise scientific notation
sich wölben to arch
wölben to arch
Wölbspiegel convex mirror
Wolfram tungsten [W, Z=74]
Würfel [M] cube
Wurfweite throwing distance
Wurzel [M] root
n-te Wurzel von x [M] nth root of x
eine Wurzel ziehen [M] to extract a root
Wurzelexponent [M] index
Xenon xenon [Xe, Z=54]
y-Achsenabschnitt [M] y-intercept
Yard [Einh.] yard
yellow yellow
Ytterbium ytterbium [Yb, Z=70]
Yttrium yttrium [Y, Z=39]
Zahl [M] number
Zahlenwert number
Zähler [M] numerator
Zählrate count rate
Zahnrad cogwheel
Zahnriemen cam belt, cambelt
Zahnriemenscheibe cam belt pulley, cam pulley
Zange pliers
Zapfen (Auge) cones (eye)
Zauberstab wand
Zedernholzöl cedar oil
Zedernöl cedar oil
Zehnerpotenz [M] power of ten
Zeichnung plan
zeigen to point
Zeit time
Zeitdilatation time dilation
Zeitkonstante time constant
Zentimeterraster centimetre grid
Zentralstrahl centre ray
Zentraltemperatur (eines Sterns) central temperature (of a star)
Zentrifugalbeschleunigung centrifugal acceleration
Zentrifugalkraft centrifugal force
Zentrifuge centrifuge
Zentripetalbeschleunigung centripetal acceleration
Zentripetalkraft centripetal force
zentrisch centric
zerbrechen to break, to crack
Zerfall decay, decomposition, disintegration
zerfallen to decay
Zerfallskonstante decay constant
Zerfallsmodus decay mode
Zerfallsreihe decay chain
zerlegen (einen Vektor) [M] to resolve (a vector)
Zerstreuungslinse diverging lens
Ziegelstein clay brick
Ziel target
Zielscheibe target
Ziffer [M] digit
Ziliarmuskel ciliary muscle
Zink zinc [Zn, Z=30]
Zinn tin [Sn, Z=50]
Zirkel compass, compasses, pair of compasses [rare]
Zirkon zirconium [Zr, Z=40]
Zitterbewegung jiggling motion
Zoll [Einh.] inch
Zugfeder extension spring
Zugfestigkeit tensile strength, ultimate tensile strength
Zugkraft pulling force
Zündkerze ignition plug, spark plug
zusammenbauen to assemble
zusammendrücken to compress
Zusammenstoß collision
Zustand state

Zustandsgrößen state variables
zweidimensionale Bewegung
two-dimensional motion
Zweig branch
Zweitaktmotor two-stroke engine
Zweitaktöl gas oil
zweiter Hauptsatz der Thermodynamik
second law of thermodynamics
zweites Kepler'sches Gesetz second Kepler law
zweites Newton'sches Gesetz second Newtonian law
Zwergplanet dwarf planet
Zwischenbild intermediate image
Zylinder [M] cylinder
Zylinder (eines Motors) cylinder (of an engine)
Zylinderarbeit work in a cylinder
Zylinderkopf cylinder head
Zylinderleistung power in a cylinder

Teil 3
Themenspezifische Wortverzeichnisse

Methods

analysis
- approximation
- on average
- to average
- average
- chart
- to conclude
- decrease
- dependence of A on B
- to depict
- diagram
- to evaluate
- evaluation
- fit
- to fit
- fitting
- graph
- increase
- to infer
- legend (of a diagram)
- nominal value
- regression line
- relation
- representation
- variation of A with B

dimension
- depth
- height
- width

experiment
- to calibrate
- calibrated
- experimental setup
- experimental set-up
- → measurement
- observation
- to optimise
- phenomena
- phenomenon
- setup

Methoden

Analyse
- Näherung
- im Mittel
- mitteln
- Mittelwert
- Messdiagramm, Messtabelle
- schlussfolgern
- Abfall, Abnahme
- Abhängigkeit von A von B
- bildlich darstellen, darstellen
- Messdiagramm
- auswerten
- Auswertung
- Anpassungskurve, Fit
- fitten
- Fitten
- Messdiagramm
- Anstieg
- schlussfolgern
- Legende (eines Diagramms)
- Sollwert
- Regressionsgerade
- Beziehung
- Darstellung
- Abhängigkeit von A von B

Dimension
- Tiefe
- Höhe
- Breite

Experiment
- kalibrieren
- kalibriert
- Versuchsaufbau
- Versuchsaufbau
- Messung
- Beobachtung
- optimieren
- Phänomene
- Phänomen
- Aufbau

Themenspezifische Wortverzeichnisse

set-up	Aufbau
to tare	tarieren
uncalibrated	unkalibriert

measurement — **Messung**

data point	Datenpunkt
to measure	messen
measured data	Messdaten, Messwerte
→ measuring device	Messgerät
to record	aufzeichnen
spreadsheet	Messtabelle
table	Messtabelle
to tabulate	tabellieren
tabulated	tabelliert

measuring device — **Messgerät**

balance	Waage
caliper	Messschieber, Schieblehre, Schublehre
calliper	Messschieber, Schieblehre, Schublehre
gauge	Messschieber, Schieblehre, Schublehre
measurement device	Messgerät
measuring instrument	Messgerät
measuring tape	Maßband
ruler	Lineal
scale	Skala, Waage

methods — **Methoden**

→ analysis	Analyse
body	Körper, Objekt
→ experiment	Experiment
generalisation	Verallgemeinerung
to investigate	untersuchen
to neglect	vernachlässigen
negligible	vernachlässigbar
object	Körper, Objekt
order of magnitude	Größenordnung
→ physical quantity	physikalische Größe
plan	Planzeichnung, Zeichnung
point-like	punktförmig
precise	genau
to question	hinterfragen
to scrutinise	hinterfragen
sketch	Skizze
special case	Spezialfall
→ theory	Theorie

Methods / Methoden

physical quantity	physikalische Größe
derived SI-unit [1]	abgeleitete SI-Einheit
fundamental constant	fundamentale Konstante
non-SI-unit [2]	Nicht-SI-Einheit
physical constant [3]	physikalische Konstante
prototype kilogram	Urkilogramm
prototype metre bar	Urmeter
scalar	Skalar
SI-base quantity [4]	SI-Basisgröße
SI-base unit [5]	SI-Grundeinheit
SI-unit [6]	SI-Einheit
vector	Vektor

proportionality	Proportionalität
constant of proportionality	Proportionalitätskonstante
to double	verdoppeln
to halve	halbieren
inversely proportional	umgekehrt proportional
inversely proportional to	umgekehrt proportional zu
proportional	proportional
proportional to	proportional zu
proportionality constant	Proportionalitätskonstante
reciprocal	Kehrwert
reciprocally proportional	umgekehrt proportional
reciprocally proportional to	umgekehrt proportional zu

scientific notation	wissenschaftliche Schreibweise
measurand	Messgröße
nomenclature	Nomenklatur
number	Zahl, Zahlenwert
power of ten	Zehnerpotenz
prefix	Präfix, Vorsilbe
prefixes	Präfixe, Vorsilben
significant figures	signifikante Stellen
symbol	Symbol
unambiguous	eindeutig
unique	eindeutig
unit	Einheit

---

[1] see chapter derived SI-units
[2] see chapter non-SI units
[3] see chapter physical constants
[4] see chapter SI-base quantities
[5] see chapter SI-base units
[6] see chapters SI-base units and derived SI-units

variable	Variable
theory	**Theorie**
approximate formula	Näherungsformel
concept	Konzept
to deduce	ableiten (logisch), herleiten
deduction	Herleitung
to derive	ableiten
to estimate	abschätzen
estimation	Abschätzung
hypotheses	Hypothesen
hypothesis	Hypothese
law of nature	Naturgesetz
model	Modell
model concept	Modellvorstellung
physics formula	physikalische Formel
presumption	Vermutung
rule of thumb	Daumenregel
miscellaneous	**Verschiedenes**
alloy	Legierung
application	Anwendung
basis	Grundlage
density	Dichte
equidistant	äquidistant
homogeneous	homogen
index	Index
indices	Indizes
to manufacture	anfertigen
manufacture	Anfertigung
material property	Materialeigenschaft
natural science	Naturwissenschaft
nature	Natur
property	Eigenschaft
property of a material	Materialeigenschaft
science	Wissenschaft
substance	Substanz
value	Wert
volume	Volumen

Kinematics

braking
- brake
- to brake
- braking distance
- braking time
- deceleration
- reaction time

circular motion
- angular frequency
- angular speed
- angular velocity
- anti-clockwise [adj.] [adv.] [B. E.]
- bend
- centrifugal acceleration
- centrifuge
- centripetal acceleration
- clockwise [adj.] [adv.]
- counterclockwise [adj.] [adv.] [A. E.]
- curve
- direction of rotation
- frequency
- to go around a bend
- to go round a curve
- hertz
- in a clockwise direction
- in a counterclockwise direction [A. E.]
- in an anti-clockwise direction [B. E.]
- inwards [adj.] [adv.]
- outwards [adj.] [adv.]
- radial
- radially inwards [adv.]
- radially outwards [adv.]
- radius
- rev counter
- revolution counter
- roller
- rotary table
- rotation
- to spin
- stroboscope
- tachometer

Kinematik

Bremsen
- Bremse
- bremsen
- Bremsweg
- Bremszeit
- Bremsverzögerung, Verzögerung
- Reaktionszeit

Kreisbewegung
- Kreisfrequenz
- Winkelgeschwindigkeit
- Vektor der Winkelgeschwindigkeit
- im Gegenuhrzeigersinn
- Kurve
- Zentrifugalbeschleunigung
- Zentrifuge
- Zentripetalbeschleunigung
- im Uhrzeigersinn
- im Gegenuhrzeigersinn
- Kurve
- Drehsinn, Rotationsrichtung
- Frequenz
- um eine Kurve fahren
- um eine Kurve fahren
- Hertz
- im Uhrzeigersinn
- im Gegenuhrzeigersinn
- im Gegenuhrzeigersinn
- nach innen
- nach außen
- radial
- radial nach innen
- radial nach außen
- Radius
- Drehzahlmesser
- Drehzahlmesser
- Walze
- Drehtisch
- Drehbewegung, Drehung, Rotation
- um die eigene Achse rotieren
- Stroboskop
- Drehzahlmesser

time period	Umlaufdauer
uniform circular motion	gleichförmige Kreisbewegung

free fall / freier Fall

acceleration due to gravity	Fallbeschleunigung
air resistance	Luftwiderstand
gravitational field strength	Fallbeschleunigung
standard acceleration due to gravity	Normwert der Fallbeschleunigung
vacuum tube	Vakuumröhre

motion / Bewegung

average speed	Durchschnittsgeschwindigkeit
average velocity	Vektor der Durchschnittsgeschwindigkeit
→ braking	Bremsen
→ circular motion	Kreisbewegung
direction of motion	Bewegungsrichtung
displacement	Abstandsvektor
distance covered	zurückgelegter Weg
→ free fall	freier Fall
initial speed	Anfangsgeschwindigkeit
instantaneous speed	Momentangeschwindigkeit
instantaneous velocity	Vektor der Momentangeschwindigkeit
kilometre per hour	Kilometer pro Stunde
light barrier	Lichtschranke
linear motion	lineare Bewegung
metre per second	Meter pro Sekunde
mileometer	Kilometerzähler
milometer [rare]	Kilometerzähler
odometer [A. E.]	Kilometerzähler
position	Ort
speed	Geschwindigkeit
speedometer	Geschwindigkeitsmesser
starting point	Anfangspunkt
stop watch	Stoppuhr
time taken	benötigte Zeit
→ two-dimensional motion	zweidimensionale Bewegung
uniform motion	gleichförmige Bewegung
uniform motion with starting point	gleichförmige Bewegung mit Anfangspunkt
→ uniformly accelerated motion	gleichmäßig beschleunigte Bewegung
velocity	Geschwindigkeitsvektor

two-dimensional motion / zweidimensionale Bewegung

flight time	Flugdauer
horizontal throw	waagrechter Wurf
maximum height	maximale Höhe

Kinematics / Kinematik

    oblique throw [A. E.]   schiefer Wurf, schräger Wurf
    path   Bahn, Flugbahn
    point of maximum height   Punkt der maximalen Höhe
    position vector   Ortsvektor
    projectile motion   schiefer Wurf, schräger Wurf
    rising time   Aufstiegszeit, Steigzeit
    throwing distance   Wurfweite
    trajectory   Trajektorie

uniformly accelerated motion   **gleichmäßig beschleunigte Bewegung**
    to accelerate   beschleunigen
    acceleration   Beschleunigung
    average acceleration   Durchschnittsbeschleunigung
    average acceleration vector   Vektor der Durchschnittsbeschleunigung
    → free fall   freier Fall
    instantaneous acceleration   Momentanbeschleunigung
    metre per second squared   Meter pro Quadratsekunde
    uniformly accelerated motion with initial speed   gleichmäßig beschleunigte Bewegung mit Anfangsgeschwindigkeit
    uniformly accelerated motion with initial speed and with starting point   gleichmäßig beschleunigte Bewegung mit Anfangsgeschwindigkeit und Anfangspunkt
    vertical throw   vertikaler Wurf

Dynamics / Dynamik

brake   Bremse
    to brake   bremsen
    brake cylinder   Bremszylinder
    brake disk   Bremsscheibe
    brake drum   Bremstrommel
    brake fluid   Bremsflüssigkeit
    brake hose   Bremsschlauch
    brake line   Bremsleitung
    brake pad   Bremsbelag
    brake pedal   Bremspedal
    brake shoe   Bremsschuh
    disk brake   Scheibenbremse
    drum brake   Trommelbremse
    handbrake   Handbremse
    return spring   Rückholfeder

deformation   Deformation, Verformung
    to compress   zusammendrücken
    to deform   deformieren, verformen

English	Deutsch
elastic deformation	elastische Deformation
elasticity	Elastizität
modelling clay	Knete, Knetmasse
modulus of elasticity	Elastizitätsmodul
plastic deformation	plastische Deformation
plasticine	Knete, Knetmasse
rubber	Gummi
rubber band	Gummiband
shape	Form
shock absorber	Stoßdämpfer
→ spring	Feder
strain	relative Längenänderung
stress	Spannung
suspension	Federung
tensile strain	relative Längenänderung
tensile strength	Zugfestigkeit
tensile stress	Spannung
ultimate tensile strength	Zugfestigkeit

fluid resistance / Strömungswiderstand

English	Deutsch
air resistance	Luftwiderstand
cross sectional area	Querschnittsfläche
drag	Strömungswiderstand
drag coefficient	Widerstandsbeiwert
dynamic pressure	Staudruck
laminar flow	laminare Strömung
slipstream	Windschatten
streamline	Stromlinie
turbulence	Turbulenz
turbulent flow	turbulente Strömung
viscosity	Viskosität

force / Kraft

English	Deutsch
balance of forces	Kräftegleichgewicht
centrifugal force	Zentrifugalkraft
centripetal force	Zentripetalkraft
contact force	Kontaktkraft
Coriolis force	Corioliskraft
counterforce	Gegenkraft
electric force	elektrische Kraft
electromagnetic force	elektromagnetische Kraft
equilibrium of forces	Kräftegleichgewicht
to exert a force	eine Kraft ausüben
→ friction	Reibung
gravitational force	Gravitationskraft

Dynamics / Dynamik

magnetic force	magnetische Kraft
newton	Newton
pseudo force	Scheinkraft
pulling force	Zugkraft
recoil	Rückstoß
resultant force	resultierende Kraft
shear force	Scherkraft
spring force	Federkraft
spring scale	Federwaage
strong force	starke Kraft
strong interaction	starke Wechselwirkung
strong nuclear force	starke Kernkraft
tension	Spannkraft
weak force	schwache Kraft
weak interaction	schwache Wechselwirkung
weak nuclear force	schwache Kernkraft
weight	Gewicht

friction / Reibung

to adhere	haften
adherence	Haftvermögen
adhesion	Adhäsion, Haftung
air bearing	Luftkissen
air bearing stage	Luftkissentisch
air cushion	Luftkissen
ball	Kugel
ball bearing	Kugellager
black ice	Glatteis
contact area	Berührungsfläche
desirable friction	erwünschte Reibung
deterioration	Abnutzung
dynamic friction	Gleitreibung
dynamic friction coefficient	Gleitreibungszahl
dynamic friction force	Gleitreibungskraft
friction coefficient	Reibungszahl
friction force	Reibungskraft
frictionless	reibungsfrei
grease	Fett
greasy	fettig, glitschig
to interlock	verzahnen
lubricant	Schmiermittel
to lubricate	schmieren
needle bearing	Nadellager
normal force	Normalkraft
normal force vector	Vektor der Normalkraft

rolling friction	Rollreibung
rolling friction coefficient	Rollreibungszahl
rolling friction force	Rollreibungskraft
roughness	Rauheit
sandpaper	Schmirgelpapier
skid marks	Bremsspur
static friction	Haftreibung
static friction coefficient	Haftreibungszahl
static friction force	Haftreibungskraft
substrate	Untergrund, Unterlage
substratum	Untergrund, Unterlage
undesirable friction	unerwünschte Reibung
wearing part	Verschleißteil
wheels lock up	Reifen blockieren
wheels spin	Räder drehen durch

inclined plane / schiefe Ebene

gradient (of an inclined plane)	Steigung (einer schiefen Ebene)
inclination (of an inclined plane)	Neigung (einer schiefen Ebene), Neigungswinkel (einer schiefen Ebene)
inclination angle (of an inclined plane)	Neigungswinkel (einer schiefen Ebene)
level plane	waagrechte Ebene
slope (of an inclined plane)	Steigung (einer schiefen Ebene)
slope angle (of an inclined plane)	Steigungswinkel (einer schiefen Ebene)

rules / Regeln

1st Newtonian law	1. Newton'sches Gesetz
2nd Newtonian law	2. Newton'sches Gesetz
3rd Newtonian law	3. Newton'sches Gesetz
actio=reactio	Actio=Reactio
first Newtonian law	erstes Newton'sches Gesetz
Hooke's law	Hooke'sches Gesetz
law of gravitation	Gravitationsgesetz
Newtonian laws	Newton'sche Gesetze
second Newtonian law	zweites Newton'sches Gesetz
Stokes' law	Stokes-Gesetz
third Newtonian law	drittes Newton'sches Gesetz

spring / Feder

to bounce	federn
coil spring	Schraubenfeder
compression spring	Druckfeder
elongation	Auslenkung, Elongation
extension spring	Zugfeder
to spring	federn
spring constant	Federkonstante

Dynamics / Dynamik

  state of motion   Bewegungszustand
- equation of motion — Bewegungsgleichung
- inertia — Trägheit
- inertial system — Inertialsystem
- to levitate — schweben, schweben lassen

  torque   Drehmoment
- moment [rare] — Drehmoment
- newton metre — Newtonmeter
- torque wrench — Drehmomentschlüssel

  miscellaneous   Verschiedenes
- end stop (e. g. of a spring scale) — Anschlag (z. B. einer Federwaage)
- hoist — Flaschenzug
- impulse — Kraftstoß
- to load — belasten
- mechanics — Mechanik
- momentum — Impuls
- pulley — Rolle, Umlenkrolle

Gravitation

  celestial object   Himmelskörper
- asteroid — Asteroid
- asteroid belt — Asteroidengürtel
- black hole — schwarzes Loch
- comet — Komet
- dwarf planet — Zwergplanet
- Kuiper belt — Kuiper-Gürtel
- moon — Mond
- neutron star — Neutronenstern
- Oort cloud — Oortsche Kometenwolke
- → planet — Planet
- planetoid — Planetoid
- remnant (of a star) — Überrest (eines Sterns)
- sun — Sonne
- supernova — Supernova
- white dwarf — weißer Zwerg

  gravitation   Gravitation, Schwerkraft
- acceleration due to gravity — Fallbeschleunigung
- to attract — sich anziehen, anziehen
- attraction — Anziehung
- equatorial surface gravity — Fallbeschleunigung an der Oberfläche am Äquator

Themenspezifische Wortverzeichnisse

English	German
gravitational constant	Gravitationskonstante
gravitational field strength	Fallbeschleunigung, Gravitationsfeldstärke
gravitational force	Schwerkraft
gravitational force vector	Vektor der Gravitationskraft
gravity	Gravitation (bezogen auf die Erde), Schwerkraft (bezogen auf die Erde)
Lagrangian point	Lagrangepunkt
microgravity	Schwerelosigkeit
Newton's law of gravitation	Newton'sches Gravitationsgesetz
standard acceleration due to gravity	Normwert der Fallbeschleunigung
surface gravity	Fallbeschleunigung an der Oberfläche
weightlessness	Schwerelosigkeit
zero gravity	Schwerelosigkeit

orbit — Umlaufbahn

English	German
eccentric	exzentrisch
eccentricity	Exzentrizität
ellipse	Ellipse
elliptic	elliptisch
geostationary	geostationär
to orbit (a planet)	umkreisen (einen Planeten)
orbit period	Umlaufdauer
orbital radius	Bahnradius
planetary motion	Planetenbewegung
planetary orbit	Planetenbahn
satellite	Satellit
satellite dish	Satellitenschüssel
semimajor axis	große Bahnhalbachse
solar distance (of a planet)	Sonnenabstand (eines Planeten)

planet — Planet

English	German
Mercury	Merkur
Venus	Venus
Earth	Erde
Mars	Mars
Jupiter	Jupiter
Saturn	Saturn
Uranus	Uranus
Neptune	Neptun
equatorial radius	Äquatorradius
globe	Globus
latitude	Breitengrad
longitude	Längengrad
mean radius	mittlerer Radius
surface temperature	Oberflächentemperatur

Gravitation / Gravitation

    volumetric mean radius   Radius der volumengleichen Kugel

rules   **Regeln**
- 1st Kepler law — 1. Kepler'sches Gesetz
- 2nd Kepler law — 2. Kepler'sches Gesetz
- 3rd Kepler law — 3. Kepler'sches Gesetz
- first Kepler law — erstes Kepler'sches Gesetz
- Kepler law — Kepler'sches Gesetz
- Newton's law of gravitation — Newton'sches Gravitationsgesetz
- second Kepler law — zweites Kepler'sches Gesetz
- third Kepler law — drittes Kepler'sches Gesetz

space   **Weltall, Weltraum**
- astronaut — Astronaut
- ion propulsion — Ionenantrieb
- to launch — starten
- manned mission — bemannte Mission
- nozzle — Düse
- outer space — ferner Weltraum, fernes Weltall
- rocket — Rakete (Raumfahrt)
- rocket nozzle — Raketendüse
- rocket stage — Raketenstufe
- space flight — Raumfahrt
- spacecraft — Raumschiff
- spaceship — Raumschiff
- universe — Universum

tides   **Gezeiten**
- high tide — Flut
- low tide — Ebbe
- tidal — Gezeiten-

miscellaneous   **Verschiedenes**
- astronomical unit — astronomische Einheit
- astronomy — Astronomie
- central temperature (of a star) — Zentraltemperatur (eines Sterns)
- light-year — Lichtjahr
- world view — Weltbild

Work, Power, Energy
Arbeit, Energie, Leistung

efficiency   Effizienz, Wirkungsgrad
- energy gain — Energieausbeute, Energiegewinn
- energy input — zugeführte Energie
- energy loss — Energieverlust
- energy output — abgegebene Energie
- energy yield — Energieausbeute, Energiegewinn
- power input — zugeführte Leistung
- power output — abgegebene Leistung
- useful energy — Nutzenergie

energy   Energie
- calorie — Kalorie
- carrier of energy — Energieträger
- to consume energy — Energie konsumieren
- energy carrier — Energieträger
- energy conservation law — Energieerhaltungssatz
- energy transfer — Energieübertragung
- energy transformation — Energieumwandlung
- energy transformation chain — Energieumwandlungskette
- heat — Wärme
- joule — Joule
- KE (kinetic energy) — kinetische Energie
- kinetic energy — kinetische Energie
- PE (potential energy) — potenzielle Energie
- potential energy — potenzielle Energie
- spring energy — Federenergie
- state — Zustand
- total energy — Gesamtenergie
- to transfer — übertragen
- to transform — umwandeln
- to use energy — Energie verbrauchen

power   Leistung
- to consume power — Leistung konsumieren
- horsepower — Pferdestärke
- to provide power — Leistung zur Verfügung stellen
- to release power — Leistung freisetzen
- watt — Watt

power station   Kraftwerk
- coal-fired power station — Kohlekraftwerk

Work, Power, Energy / Arbeit, Energie, Leistung

    fossil   fossil
    fossil fuel   fossiler Brennstoff, fossiler Energieträger
    geothermal power station   geothermisches Kraftwerk
    hydroelectric power station   Wasserkraftwerk
    nuclear plant   Kernkraftwerk
    nuclear power station   Kernkraftwerk
    power plant   Kraftwerk
    solar power station   Solarkraftwerk
    wind power station   Windkraftwerk
    wind-driven power station   Windkraftwerk

system   System
    closed system   abgeschlossenes System
    open system   offenes System

work   Arbeit
    acceleration work   Beschleunigungsarbeit
    to do work   Arbeit verrichten
    friction work   Reibungsarbeit
    lifting work   Hubarbeit
    to perform work   Arbeit verrichten
    process   Vorgang
    work performed on a spring   Federarbeit

Hydrostatics / Hydrostatik

air pressure   Luftdruck
    altimeter   Höhenmesser
    aneroid barometer   Dosenbarometer
    aneroid capsule (of an aneroid barometer)   Vidi-Dose (eines Dosenbarometers)
    aneroid cell (of an aneroid barometer)   Vidi-Dose (eines Dosenbarometers)
    atmosphere (e. g. of Earth)   Atmosphäre (z. B. der Erde)
    barometer   Barometer
    column of mercury   Quecksilbersäule
    Magdeburg sphere   Magdeburger Kugel
    millimetre of mercury   Millimeter Quecksilbersäule
    outer atmosphere   äußere Atmosphäre
    sea level   Meereshöhe
    standard pressure   Normdruck

buoyancy   Auftrieb, Auftriebskraft
    aerometer   Aerometer
    to displace   verdrängen

to displace water	Wasser verdrängen
displaced	verdrängt
displaced liquid	verdrängte Flüssigkeit
to float	schweben (in der Luft), schweben (unter Wasser), schwimmen (Objekte oder Lebewesen, die auf dem Wasser treiben), treiben (Objekte oder Lebewesen auf dem Wasser)
to immerse	eintauchen
to rise (e. g. in water)	aufsteigen (z. B. in Wasser)
to sink (e. g. in water)	untergehen (z. B. im Wasser)
to swim	schwimmen (Lebewesen auf dem Wasser durch eigenen Krafteinsatz)
upthrust [rare]	Auftrieb, Auftriebskraft
water displacement	Wasserverdrängung

diving — Tauchen

- diver — Taucher
- lung — Lunge
- scuba (self-contained underwater breathing apparatus) — Tauchgerät
- scuba diving — Gerätetauchen

hydraulics — Hydraulik

- hydraulic — hydraulisch
- hydraulic lift — Hebebühne
- hydraulic press — hydraulische Presse

phase — Aggregatzustand, Phase

- boiling point — Siedepunkt
- condensation — Kondensation
- to condense — kondensieren
- to deposit — resublimieren
- deposition — Resublimation
- fluid — Fluid
- to freeze — erstarren
- freezing — Gefrieren
- freezing point — Gefrierpunkt
- gas — Gas
- gaseous — gasförmig
- liquid — flüssig, Flüssigkeit
- to melt — schmelzen
- melting — Schmelzen
- melting point — Schmelzpunkt
- molten (e. g. iron) — geschmolzen (z. B. Eisen)
- phase transition — Phasenübergang

Hydrostatics / Hydrostatik

solid	fest, Festkörper
to sublimate	sublimieren
sublimation	Sublimation
vaporisation	Verdampfen
to vaporise	verdampfen
vapour	Dampf

pneumatics — Pneumatik

air pump	Luftpumpe
bicycle pump	Fahrradpumpe
hovercraft	Luftkissenschiff
to inflate	aufblasen
pneumatic	pneumatisch
pneumatic hammer	Presslufthammer
pneumatic-tyred	luftbereift
tyre	Reifen
tyre pump	Luftpumpe
valve	Ventil

pressure — Druck

→ air pressure	Luftdruck
atmosphere	Atmosphäre
bar	Bar
Boyle-Mariotte experiment	Boyle-Mariotte-Experiment
communicating vessels	kommunizierende Röhren
compressed air	Druckluft
compressibility	Kompressibilität
to exert pressure	Druck ausüben
hose	Schlauch (groß)
hose pressure	Schlauchdruck
hydrostatic paradox	hydrostatisches Paradoxon
hydrostatic pressure	Schweredruck
incompressibility	Inkompressibilität
manometer	Manometer
overpressure	Überdruck
pascal	Pascal
pump	Pumpe
total pressure	Gesamtdruck
underpressure [rare]	Unterdruck
U-tube	U-Rohr
U-tube manometer	U-Manometer
vacuum	Unterdruck, Vakuum

rules — Regeln

barometric formula	barometrische Höhenformel
Boyle-Mariotte's law	Boyle-Mariotte'sches Gesetz

law of Archimedes	Gesetz des Archimedes
Pascal's law	Gesetz von Pascal

water | Wasser
- aquarium — Aquarium
- dam — Staudamm
- distilled water — destilliertes Wasser
- fresh water — Süßwasser
- iceberg — Eisberg
- moisture — Feuchtigkeit
- salt water — Salzwasser
- sea water — Meerwasser, Salzwasser
- water jet — Wasserstrahl
- water reservoir — Wasserreservoir
- water vapour — Wasserdampf

Heat / Wärme

combustion | Verbrennung
- calorific value — Brennwert, Heizwert
- combustion chamber — Brennkammer
- gross calorific value — Brennwert, oberer Heizwert [alt]
- gross heat of combustion — Brennwert, oberer Heizwert [alt]
- heat of combustion — Brennwert, Heizwert
- heating value — Brennwert, Heizwert
- higher heating value — Brennwert, oberer Heizwert [alt]
- lower heating value — Heizwert, unterer Heizwert [alt]
- net calorific value — Heizwert, unterer Heizwert [alt]
- net heat of combustion — Heizwert, unterer Heizwert [alt]

convection | Konvektion, Wärmeströmung
- air circulation — Luftzirkulation
- convection tube — Konvektionsrohr
- great ocean conveyor belt — Globales Förderband
- Gulf Stream — Golfstrom

electromagnetic spectrum | elektromagnetisches Spektrum
- infrared light — infrarotes Licht
- ultraviolet light — ultraviolettes Licht
- visible light — sichtbares Licht

equilibrium processes in gases | Gleichgewichtsprozesse in Gasen
- adiabatic — adiabatisch
- adiabatic coefficient — Adiabatenkoeffizient
- isobaric — isobar

Heat / Wärme

    isochoric   isochor
    isothermal   isotherm

heat conduction   **Wärmeleitung**
    to conduct (thermally)   leiten (thermisch)
    conduction (thermal)   Leitung (thermisch)
    conductor (thermal)   Leiter (thermisch)
    heat conduction coefficient   Wärmeleitfähigkeitskoeffizient

heat engine   **Wärmekraftmaschine**
    axle   Achse
    belt   Riemen
    cam   Nocke
    cam belt   Zahnriemen
    cam belt pulley   Zahnriemenscheibe
    cam pulley   Zahnriemenscheibe
    cambelt   Zahnriemen
    camshaft   Nockenwelle
    carburetor [A. E.]   Vergaser
    carburettor [B. E.]   Vergaser
    chain   Kette
    clutch   Kupplung
    clutch disk   Kupplungsscheibe
    cogwheel   Zahnrad
    compression (stroke 2 of a four-stroke engine)   Verdichten (Takt 2 eines Viertaktmotors)
    crank   Kurbel
    to crank   kurbeln
    crankshaft   Kurbelwelle
    cylinder (of an engine)   Zylinder (eines Motors)
    cylinder capacity   Hubraum
    cylinder head   Zylinderkopf
    engine   Motor
    exhaust (stroke 4 of a four-stroke engine)   Ausstoßen (Takt 4 eines Viertaktmotors)
    expansion   Expansion
    explosion   Explosion
    external combustion   externe Verbrennung
    four-stroke engine   Viertaktmotor
    heat pump   Wärmepumpe
    ignition plug   Zündkerze
    intake (stroke 1 of a four-stroke engine)   Ansaugen (Takt 1 eines Viertaktmotors)
    internal combustion   interne Verbrennung
    maximum efficiency   maximaler Wirkungsgrad
    O-ring   O-Ring

perpetuum mobile	Perpetuum mobile
perpetuum motion machine	Perpetuum mobile
piston	Kolben
piston rod	Pleuel, Pleuelstange
power (stroke 3 of a four-stroke engine)	Arbeiten (Takt 3 eines Viertaktmotors)
power in a cylinder	Zylinderleistung
pulley	Riemenscheibe
spark plug	Zündkerze
steam engine	Dampfmaschine
stroke	Takt (Motor)
two-stroke engine	Zweitaktmotor
valve	Ventil
V-belt	Keilriemen
work in a cylinder	Zylinderarbeit

heat propagation / Wärmeausbreitung

ablation	Ablation
→ convection	Konvektion
→ heat conduction	Wärmeleitung
→ insulation (thermal)	Isolation (thermisch)
→ thermal radiation	Wärmestrahlung

insulation (thermal) / Isolation (thermisch)

double glazing	Doppelverglasung
double-walled	doppelwandig
heat bridge	Wärmebrücke
heat transfer	Wärmedurchgang
heat transfer coefficient	Wärmedurchgangskoeffizient
to insulate (thermally)	isolieren (thermisch)
insulator (thermal)	Isolator (thermisch)
thermos flask	Thermosflasche
window pane	Fensterscheibe

internal energy / innere Energie

heat capacity	Wärmekapazität
latent heat	latente Wärme
specific heat	spezifische Wärme
specific heat capacity	spezifische Wärmekapazität
specific latent heat of fusion	spezifische Schmelzwärme
specific latent heat of vaporization	spezifische Verdampfungswärme

lab equipment / Laborausrüstung

bar	Stab, Stange
base (of a stand)	Sockel (eines Stativs)
beaker	Becherglas
burner	Brenner

Heat / Wärme

Erlenmeyer flask	Erlenmeyerkolben
glass tube	Glasrohr
immersion heater	Tauchsieder
pipe	Rohr, Schlauch
plug	Stopfen
pole	Stab, Stange
rod	Stab, Stange
stand	Stativ
test tube	Reagenzglas
tube	Rohr, Schlauch

lime — **Kalk**

to fur up	verkalken
limestone	Kalkstein

particle model — **Teilchenmodell**

atomic mass unit	atomare Masseneinheit
Avogadro's number	Avogadro-Zahl
Boltzmann constant	Boltzmann-Konstante
fat droplet	Fetttröpfchen
gas constant	allgemeine Gaskonstante
gas particle	Gasteilchen
jiggling motion	Zitterbewegung
kinetic energy of a gas particle	kinetische Energie eines Gasteilchens
macrocosm	Makrokosmos
macroscopic	makroskopisch
microcosm	Mikrokosmos
microscopic	mikroskopisch
particle	Teilchen
particle number	Teilchenzahl
particle number density	Teilchenzahldichte

rules — **Regeln**

1st law of thermodynamics	1. Hauptsatz der Thermodynamik
2nd law of thermodynamics	2. Hauptsatz der Thermodynamik
3rd law of thermodynamics	3. Hauptsatz der Thermodynamik
Amonton's law	Amonton'sches Gesetz
Boyle-Mariotte's law	Boyle-Mariotte'sches Gesetz
first law of thermodynamics	erster Hauptsatz der Thermodynamik
Gay-Lussac's law	Gay-Lussac'sches Gesetz
ideal gas equation	allgemeine Gasgleichung
second law of thermodynamics	zweiter Hauptsatz der Thermodynamik
Stefan-Boltzmann law	Stefan-Boltzmann-Gesetz
third law of thermodynamics	dritter Hauptsatz der Thermodynamik

English	Deutsch
standard conditions	Normbedingungen
standard acceleration due to gravity	Normwert der Fallbeschleunigung
standard pressure	Normdruck
standard state	Normzustand
standard temperature	Normtemperatur
standard value	Normwert
temperature	Temperatur
absolute temperature	absolute Temperatur
absolute zero	absoluter Nullpunkt
Celsius scale	Celsius-Skala
Celsius temperature	Celsius-Temperatur
degree Celsius	Grad Celsius
degree Fahrenheit	Grad Fahrenheit
Fahrenheit scale	Fahrenheit-Skala
Fahrenheit temperature	Fahrenheit-Temperatur
kelvin	Kelvin
temperature of the mixture	Mischtemperatur
thermal expansion	Wärmeausdehnung
linear thermal expansion	Längenausdehnung
linear thermal expansion coefficient	Längenausdehnungskoeffizient
thermal expansion coefficient	Wärmeausdehnungskoeffizient
volumetric thermal expansion	Volumenausdehnung
volumetric thermal expansion coefficient	Volumenausdehnungskoeffizient
thermal radiation	Wärmestrahlung
absorption	Absorption
emission	Emission
emissivity	Emissivität
to emit	emittieren
to glow	glühen
matt	matt
metallized	verspiegelt
mirrored	verspiegelt
parabolic mirror	Parabolspiegel
radiator	Heizkörper
to reflect	reflektieren
shiny	glänzend
Stefan-Boltzmann constant	Stefan-Boltzmann-Konstante
thermal imaging camera	Wärmebildkamera
thermography	Thermographie
thermodynamic equilibrium	thermodynamisches Gleichgewicht
→ equilibrium processes in gases	Gleichgewichtsprozesse in Gasen

Heat / Wärme

    state variables   Zustandsgrößen
    thermodynamics   Thermodynamik, Wärmelehre

thermometer — Thermometer
    bimetallic thermometer   Bimetallthermometer
    liquid thermometer   Flüssigkeitsthermometer
    mercury thermometer   Quecksilberthermometer
    resistance thermometer   Widerstandsthermometer
    temperature sensor   Temperaturfühler
    thermochrome lacquer   Thermochromlack
    thermochrome paper   Thermochrompapier

Electrostatics — Elektrostatik

applications — Anwendungen
    accelerator   Beschleuniger
    cathode ray tube   Kathodenstrahlrohr
    linear accelerator   Linearbeschleuniger
    particle accelerator   Teilchenbeschleuniger

atom[7] — Atom
    atomic shell   Atomhülle, Atomschale
    atomic structure   Atombau, Atomstruktur
    electron   Elektron

atomic nucleus[8] — Atomkern
    atomic nuclei   Atomkerne
    neutron   Neutron
    nuclei   Atomkerne, Kerne
    nucleus   Atomkern, Kern
    proton   Proton

attraction — Anziehung
    to attract   sich anziehen, anziehen
    to attract each other   sich gegenseitig anziehen
    attractive   anziehend

capacitor — Kondensator
    capacitance   Kapazität (eines Kondensators)
    capacitor plate   Kondensatorplatte
    energy density   Energiedichte
    farad   Farad

---

[7] see also chapter quantum physics
[8] see also chapter radioactivity

parallel plate capacitor	Plattenkondensator
plate area	Plattenfläche
plate distance	Plattenabstand
variable capacitor	Drehkondensator
volt	Volt
voltage	Spannung
wound capacitor	Wickelkondensator

charge — Ladung

atomic unit of charge	atomare Ladungseinheit
to charge	aufladen, laden
charging	Aufladen, Aufladung
charging-up	Aufladung
coulomb	Coulomb
to discharge	entladen
discharge	Entladung
electric charge	elektrische Ladung
electricity	Elektrizität
electroscope	Elektroskop
electrostatic induction	Influenz
elementary charge	Elementarladung
glow lamp	Glimmlampe
glow-discharge lamp	Glimmlampe
lightning	Blitz
lightning conductor	Blitzableiter
negatively charged	negativ geladen
opposite sign	entgegengesetztes Vorzeichen
point charge	Punktladung
positively charged	positiv geladen
ribbon	Band
ribbon generator	Bandgenerator
to rub	reiben
same sign	dasselbe Vorzeichen
test charge	Probeladung
van de Graaff generator	Van de Graaff Generator
wand	Zauberstab

chemical bond[9] — chemische Bindung

atomic lattice	Atomgitter
bound electron	gebundenes Elektron
combined with the lattice	gebunden an das Gitter
covalent bond	kovalente Bindung
free electron	freies Elektron

---

[9] see also chapter chemistry

Electrostatics / Elektrostatik

    ionic bond — Ionenbindung
    lattice — Gitter (eines Festkörpers), Kristallgitter
    metallic bond — Metallbindung
    salt crystal — Salzkristall
    table salt crystal — Kochsalzkristall

Coulomb's law / Coulomb-Gesetz

    Coulomb force — Coulombkraft
    Coulomb force vector — Vektor der Coulombkraft
    electric constant — Dielektrizitätskonstante des Vakuums [alt], elektrische Feldkonstante, Permittivität des Vakuums [selten]
    electric field constant [rare] — Dielektrizitätskonstante des Vakuums [alt], elektrische Feldkonstante, Permittivität des Vakuums [selten]
    permittivity of free space — Dielektrizitätskonstante des Vakuums [alt], elektrische Feldkonstante, Permittivität des Vakuums [selten]
    permittivity of vacuum — Dielektrizitätskonstante des Vakuums [alt], elektrische Feldkonstante, Permittivität des Vakuums [selten]

dielectrics / Dielektrika

    dielectric — Dielektrikum
    relative permittivity — Dielektrizitätszahl

electric field / elektrisches Feld

    castor oil — Rizinusöl
    counter field — Gegenfeld
    electric field line — elektrische Feldlinie
    electric field strength — elektrische Feldstärke
    Faraday cage — Faraday-Käfig
    field strength — Feldstärke
    homogenous — homogen
    semolina grain — Grieskorn
    to visualise (field lines) — sichtbar machen (Feldlinien)

pole / Pol

    dipole — Dipol
    like poles — gleichnamige Pole
    minus pole — Minuspol
    negative pole — negativer Pol
    neutral — neutral
    plus pole — Pluspol
    polar — polar
    polarisation — Polarisation

polarity — Polarität
positive pole — positiver Pol
unlike poles — ungleichnamige Pole

repulsion — Abstoßung
to repel — abstoßen
to repel each other — sich gegenseitig abstoßen
repulsive — abstoßend

Electrodynamics / Elektrodynamik

battery — Batterie
capacity (of a battery) — Kapazität (einer Batterie)
flat battery — leere Batterie

earthing [B. E.] — Erden, Erdung
earth (electric wire) [B. E.] — Erde (elektrisches Kabel)
to earth [B. E.] — erden
earthed [B. E.] — geerdet
ground (electric wire) [A. E.] — Erde (elektrisches Kabel)
to ground [A. E.] — erden
ground [A. E.] — Erdung
ground wire [A. E.] — Erde (elektrisches Kabel)
grounded [A. E.] — geerdet
grounding [A. E.] — Erden, Erdung

electric circuit — elektrischer Stromkreis
banana plug — Bananenstecker
circuit diagram — Schaltbild, Schaltplan
circuit symbol — Schaltsymbol
closed circuit — geschlossener Stromkreis
closed electric circuit — geschlossener Stromkreis
connected in parallel — parallel geschaltet
connected in series — in Reihe geschaltet, seriell geschaltet
conventional current direction — konventionelle Stromrichtung, technische Stromrichtung

crocodile clamp — Krokodilklemme
diode — Diode
electrical component — elektrisches Bauteil
germanium diode — Germaniumdiode
in parallel — in Parallel
parallel circuit — Parallelschaltung
parallel connection — Parallelschaltung
plug — Stecker
pre-resistor — Vorwiderstand

Electrodynamics / Elektrodynamik

real current direction	physikalische Stromrichtung
resistor	Widerstand (Bauteil)
resistor combination	Widerstandskombination
in series	in Reihe
series circuit	Reihenschaltung
series connection	Reihenschaltung
short	Kurzschluss
short circuit	Kurzschluss
to short-circuit	kurzschließen
short-circuit	Kurzschluss
switch	Schalter
toggle	Umschalter
wire	Draht

electric current — elektrische Stromstärke, elektrischer Strom

AC	Wechselstrom
AC current	Wechselstrom
alternating current	Wechselstrom
ammeter	Amperemeter
amp	Ampere
ampere	Ampere
amperemeter	Amperemeter
current	Strom
DC	Gleichstrom
DC current	Gleichstrom
direct current	Gleichstrom

electric shock — elektrischer Schlag, Stromschlag

cardiac fibrillation	Herzflimmern
cramp	Krampf
danger of death	Todesgefahr
heartbeat	Herzschlag
internal resistance of the human body	Innenwiderstand des menschlichen Körpers
skin resistance	Übergangswiderstand der Haut

electrical appliance — elektrisches Gerät

cooker [B. E.]	Herd
electric radiator	elektrischer Heizkörper
heating wire	Heizdraht
hotplate	Herdplatte
housing	Gehäuse
iron	Bügeleisen
oven	Ofen

quartz watch	Quarzuhr
radiator	Heizkörper
stove	Herd
tram	Tram

electrical resistance — elektrischer Widerstand (physikalische Größe)

characteristic	Kennlinie
characteristic curve	Kennlinie
to conduct (electrically)	leiten (elektrisch)
conduction (electrical)	Leitung (elektrisch)
conductor (electrical)	Leiter (elektrisch)
current-voltage characteristic	Strom-Spannungs-Kennlinie
current-voltage characteristic curve	Strom-Spannungs-Kennlinie
dimmer	Dimmer
electric resistance	elektrischer Widerstand (physikalische Größe)
electric resistor	elektrischer Widerstand (Bauteil)
electrical resistor	elektrischer Widerstand (Bauteil)
to insulate (electrically)	isolieren (elektrisch)
insulation (electrical)	Isolation (elektrisch)
insulator (electrical)	Isolator (elektrisch)
internal resistance	Innenwiderstand
leakage current	Leckstrom
metallic conductor	metallischer Leiter
ohm	Ohm
→ ohmic resistor	ohmscher Widerstand (das elektrische Bauteil)
ohmmeter	Ohmmeter
potentiometer	Potenziometer
resistance	Widerstand (physikalische Größe)
resistivity	spezifischer Widerstand
semiconductor	Halbleiter
specific resistance	spezifischer Widerstand
superconductor	Supraleiter
variable resistor	Schiebewiderstand

electrode — Elektrode

anode	Anode
cathode	Kathode
glow cathode	Glühkathode

fuse — Sicherung

ALCI (appliance leakage current interrupter)	Fehlerstromschutzschalter, FI-Schutzschalter, FI-Sicherung

Electrodynamics / Elektrodynamik

appliance leakage current interrupter	Fehlerstromschutzschalter, FI-Schutzschalter, FI-Sicherung
a fuse blows	eine Sicherung fliegt heraus
a fuse is blown	eine Sicherung ist herausgeflogen
GFCI (ground fault circuit interrupter)	Fehlerstromschutzschalter, FI-Schutzschalter, FI-Sicherung
ground fault circuit interrupter	Fehlerstromschutzschalter, FI-Schutzschalter, FI-Sicherung
RCD (residual current device)	Fehlerstromschutzschalter, FI-Schutzschalter, FI-Sicherung
residual current device	Fehlerstromschutzschalter, FI-Schutzschalter, FI-Sicherung
safety fuse	Sicherung

grid [B. E.] — Stromnetz
- European integrated network — Europäisches Verbundnetz
- kilowatt-hour — Kilowattstunde
- mains — Stromnetz
- outlet — Steckdose
- power outlet — Steckdose
- power point — Steckdose
- socket — Steckdose
- → three wire cable — dreiadriges Kabel
- transformer — Transformator
- transmission line — Hochspannungsleitung
- to transmit — übermitteln

ionisation — Ionisation
- anion — Anion
- cation — Kation
- ion — Ion

Kirchhoff's circuit laws — Kirchhoffsche Gesetze
- branch — Zweig
- KCL (Kirchhoff's current law) — Kirchhoffsche Knotenregel, Knotenregel
- Kirchhoff's current law — Kirchhoffsche Knotenregel, Knotenregel
- Kirchhoff's voltage law — Kirchhoffsche Maschenregel, Maschenregel
- KVL (Kirchhoff's voltage law) — Kirchhoffsche Maschenregel, Maschenregel
- loop — Schleife
- node — Knoten

lamp — Lampe
- energy-saving bulb — Sparlampe
- filament (of a light bulb) — Glühfaden (einer Glühbirne)

fluorescent lamp	Leuchtstoffröhre
halogen lamp	Halogenlampe
head light (car)	Scheinwerfer (Auto)
incandescent bulb	Glühbirne
incandescent light bulb	Glühbirne
LED (light emitting diode)	LED (lichtemittierende Diode), Leuchtdiode
light bulb	Glühbirne
light emitting diode	Leuchtdiode
to shine (e. g. a light bulb)	leuchten (z. B. eine Glühbirne)
torch	Taschenlampe

ohmic resistor / ohmscher Widerstand (das elektrische Bauteil)

colour code	Widerstandscode
encrypted	verschlüsselt
international colour code	internationaler Farbcode, internationaler Widerstandscode
Ohm's law	Ohm'sches Gesetz
ohmic resistance	ohmscher Widerstand (die Eigenschaft)

three wire cable / dreiadriges Kabel

earth wire [B. E.]	Erdkabel
neutral wire	Nullleiter (das Kabel)
phase (the phase wire)	Phase (das elektrische Kabel)
phase wire	Phase (das elektrische Kabel)

voltage / Spannung

AC voltage	Wechselspannung
alternating voltage	Wechselspannung
to apply voltage	Spannung anlegen
to apply voltage to a resistor	Spannung an einen Widerstand anlegen
DC voltage	Gleichspannung
direct current voltage	Gleichspannung
direct voltage	Gleichspannung
high voltage	Hochspannung
voltage drop	Spannungsabfall
voltage drops at a resistor	Spannung fällt an einem Widerstand ab
voltage source	Spannungsquelle
voltages in the parallel circuit	Spannungen bei der Parallelschaltung
voltages in the series circuit	Spannungen bei der Reihenschaltung
voltmeter	Voltmeter

miscellaneous / Verschiedenes

constantan	Konstantan
course (e. g. of a voltage)	Verlauf (z. B. einer Spannung)

Electrodynamics / Elektrodynamik

current divider	Stromteiler
electric power	elektrische Leistung
electrical work	Arbeit des elektrischen Stroms
electron density	Elektronendichte
graphite	Graphit
potential divider	Spannungsteiler
power at a resistor	Leistung in einem Widerstand
power loss	Leistungsverlust
voltage divider	Spannungsteiler

Magnetism and Electric Motor
Magnetismus und Elektromotor

applications — Anwendungen
- accelerator ring — Kreisbeschleuniger, Ringbeschleuniger
- → fine beam tube — Fadenstrahlrohr
- → Hall probe — Hall-Sonde
- mass spectrometer — Massenspektrometer
- particle accelerator ring — Teilchenbeschleunigungsring
- Wehnelt tube — Wehneltröhre

electric motor — Elektromotor
- AC motor — Wechselstrommotor
- armature — Rotor
- brush (motor, generator) — Bürste (Motor, Generator)
- commutator — Kommutator
- DC motor — Gleichstrommotor
- direction of rotation — Drehsinn, Rotationsrichtung
- drive belt — Antriebsriemen
- drive pulley — Antriebsriemenscheibe, Antriebsscheibe
- half ring — Halbring
- motor — Motor
- rotor — Rotor
- sliding contact — Gleitkontakt
- stator — Stator
- stop position — Halteposition, Ruheposition
- synchronous motor — Synchronmotor
- universal motor — Universalmotor

electromagnetism — Elektromagnetismus
- circular current — Kreisstrom
- coil — Spule
- coil length — Spulenlänge
- conducting wire — Leiterdraht

conductor swing	Leiterschaukel
electromotive force	Biot-Savart-Kraft
length of a coil	Spulenlänge
loop (of a coil)	Windung (einer Spule)
magnetic core	Eisenkern
number of loops	Windungszahl
solenoid	Spule (sehr groß)
wire loop	Leiterschlaufe

fine beam tube — Fadenstrahlrohr

acceleration unit	Beschleunigungseinheit
charge-to-mass ratio for the electron	Ladungs-Massenverhältnis des Elektrons
circular path	Kreisbahn
collimation	Kollimation
counter-voltage	Gegenspannung
glow cathode	Glühkathode
helical	schraubenförmig
helical path	Schraubenbahn
helix	Schraubenlinie
Helmholtz coils	Helmholtz-Spulen
Lorentz force	Lorentzkraft
thermal electrons	thermische Elektronen
trajectory	Trajektorie

geomagnetism — Erdmagnetismus, Geomagnetismus

aurora	Polarlicht
aurora australis	südliches Polarlicht, Südlicht
aurora borealis	nördliches Polarlicht, Nordlicht
compass	Kompass
compass needle	Kompassnadel
declination	Deklination
Earth's magnetic field	Erdmagnetfeld
inclination	Inklination
North Pole	Nordpol (der Erde)
northern lights	nördliches Polarlicht, Nordlicht
radiation belt	Strahlungsgürtel
solar wind	Sonnenwind
South Pole	Südpol (der Erde)
southern lights	südliches Polarlicht, Südlicht
Van Allen belt	Van-Allen-Gürtel

Hall probe — Hall-Sonde

electron density	Elektronendichte
Hall effect	Hall-Effekt

magnetic field / Magnetfeld

English	German
homogeneous	homogen
iron filings	Eisenfeilspäne, Eisenspäne
magnetic field axis	Achse des magnetischen Feldes
magnetic field line	Magnetfeldlinie, magnetische Feldlinie
magnetic field strength	Magnetfeldstärke, magnetische Feldstärke
magnetic flux	magnetischer Fluss
magnetic flux density	magnetische Flussdichte
north pole	Nordpol (eines Magneten)
permeability of vacuum	magnetische Feldkonstante
relative permeability	Permeabilitätszahl
south pole	Südpol (eines Magneten)
tesla	Tesla
to visualise (field lines)	sichtbar machen (Feldlinien)
weber	Weber

magnetic response / magnetische Reaktion

English	German
diamagnet	Diamagnet
diamagnetic	diamagnetisch
diamagnetism	Diamagnetismus
ferromagnet	Ferromagnet
ferromagnetic	ferromagnetisch
ferromagnetism	Ferromagnetismus
magnetic induction	magnetische Influenz
magnetisation	Magnetisierung
to magnetise	magnetisieren
magnetised	magnetisiert
paramagnet	Paramagnet
paramagnetic	paramagnetisch
paramagnetic substance	paramagnetische Substanz
paramagnetism	Paramagnetismus

permanent magnetism / Permanentmagnetismus

English	German
bar magnet	Stabmagnet
basic magnet	Elementarmagnet
Curie temperature	Curietemperatur
horseshoe magnet	Hufeisenmagnet
magnetite	Magnetit
permanent magnet	Permanentmagnet

rules / Regeln

English	German
cause, transmission, effect	Ursache, Vermittlung, Wirkung
right hand rule	Rechte-Hand-Regel
three finger rule	Dreifingerregel
three finger rule of the right hand	Dreifingerregel der rechten Hand

Induction and Generator
Induktion und Generator

electromagnetic induction | Induktion
- eddy current — Wirbelstrom
- eddy current brake — Wirbelstrombremse
- henry — Henry
- induced voltage — induzierte Spannung
- inductance — Induktivität
- induction — Induktion
- induction cooker [B. E.] — Induktionsherd
- induction stove — Induktionsherd
- induction voltage — Induktionsspannung

generator | Generator
- armature — Rotor
- bicycle dynamo — Fahrraddynamo
- brush (motor, generator) — Bürste (Motor, Generator)
- commutator — Kommutator
- drive belt — Antriebsriemen
- drive pulley — Antriebsriemenscheibe, Antriebsscheibe
- dynamo — Dynamo, Lichtmaschine (Auto)
- dynamo-electric principle — dynamoelektrisches Prinzip
- generator current — Generatorstrom
- generator voltage — Generatorspannung
- half ring — Halbring
- power generator — Kraftwerksgenerator
- rotor — Rotor
- sliding contact — Gleitkontakt
- stator — Stator
- stop position — Halteposition, Ruheposition
- universal generator — Universalgenerator

rules | Regeln
- cause, transmission, effect — Ursache, Vermittlung, Wirkung
- Lenz's rule — Lenz'sche Regel
- right hand rule — Rechte-Hand-Regel
- three finger rule — Dreifingerregel
- three finger rule of the right hand — Dreifingerregel der rechten Hand

transformer | Transformator
- energy flow — Energiefluss
- primary circuit — Primärkreis
- primary coil — Primärspule

Induction and Generator / Induktion und Generator

secondary circuit	Sekundärkreis
secondary coil	Sekundärspule
transformer current relation	Stromstärkebeziehung beim Transformator
transformer equation	Transformatorengleichung
transformer power relation	Leistungsbeziehung beim Transformator
transformer station	Transformatorenstation
transformer voltage relation	Spannungsbeziehung beim Transformator

Oscillations / Schwingungen

clock	Uhr
balance wheel	Unruh (einer Uhr)
pendulum clock	Pendeluhr
spring pendulum mechanism	Federpendelmechanismus
damping	Dämpfung
to attenuate	abklingen, sich abschwächen
damped oscillation	gedämpfte Schwingung
damping coefficient	Dämpfungsfaktor
to decay (oscillation)	abklingen, sich abschwächen
heavy damping	starke Dämpfung
light damping	schwache Dämpfung
lightly damped oscillation	schwach gedämpfte Schwingung
time constant	Zeitkonstante
undamped oscillation	ungedämpfte Schwingung
viscous damping coefficient	viskoser Dämpfungsfaktor
forced oscillation	erzwungene Schwingung
eigenfrequency	Eigenfrequenz
to excite (oscillation)	anregen (Schwingung)
exciter (of an oscillation)	Erreger (einer Schwingung)
natural frequency	Eigenfrequenz
resonance	Resonanz
resonance disaster	Resonanzkatastrophe
resonant frequency	Resonanzfrequenz
resonant value of the amplitude	Resonanzamplitude
vibration generator	Schwingungsgenerator
oscillation	Schwingung
acceleration amplitude	Beschleunigungsamplitude
acceleration-elongation relation	Beschleunigungs-Elongationsbeziehung
amplitude	Amplitude
angular frequency	Kreisfrequenz
angular speed	Winkelgeschwindigkeit

angular velocity	Vektor der Winkelgeschwindigkeit
deflection	Auslenkung, Elongation
disturbance	Störung
elongation	Auslenkung, Elongation
equilibrium	Gleichgewicht
equilibrium position	Gleichgewichtslage
frequency	Frequenz
harmonic oscillation	harmonische Schwingung
→ pendulum	Pendel
period	Schwingungsdauer
periodic	periodisch
→ phase (of an oscillation)	Phase (einer Schwingung)
projection	Projektion
rest position	Ruhelage
restoring force	rücktreibende Kraft
SHM (simple harmonic motion)	harmonische Schwingung
simple harmonic motion	harmonische Schwingung
speed amplitude	Geschwindigkeitsamplitude
vibratory system	schwingungsfähiges System
zero-crossing	Nulldurchgang

pendulum — Pendel

English	German
bob	Pendelkörper
pendulum bob	Pendelkörper
simple pendulum	Fadenpendel
spring pendulum	Federpendel
string pendulum	Fadenpendel

phase (of an oscillation) — Phase (einer Schwingung)

English	German
in phase	in Phase
opposite in phase	gegenphasig
phase-shifted	phasenverschoben

superposition — Superposition, Überlagerung

English	German
beat	Schwebung
beat frequency	Schwebungsfrequenz
constructive interference	konstruktive Interferenz
destructive interference	destruktive Interferenz
interference	Interferenz
modulated	moduliert
resulting oscillation	resultierende Schwingung

Waves

coupling
    coupled oscillators
    longitudinal coupling
    transverse coupling

diffraction
    diffraction angle
    diffraction grating
    double slit
    grating
    main maximum
    maxima
    maximum
    minima
    minimum
    order of the main maximum
    order of the maximum
    pin hole
    pinhole
    single slit

disturbance
    compression disturbance
    rarefaction disturbance

earthquake
    Earth's core
    Earth's crust
    Earth's mantle
    epicentre
    hypocentre
    primary wave
    secondary wave
    seismic wave
    seismogram

Fourier analysis
    frequency component
    fundamental
    fundamental oscillation
    fundamental tone
    harmonic
    overtone

Wellen

Kopplung
    gekoppelte Oszillatoren
    longitudinale Kopplung
    transversale Kopplung

Beugung
    Beugungswinkel
    Beugungsgitter, Gitter
    Doppelspalt
    Beugungsgitter, Gitter
    Hauptmaximum
    Maxima
    Maximum
    Minima
    Minimum
    Ordnung des Hauptmaximums
    Ordnung des Maximums
    Nadelloch, Stiftloch
    Nadelloch, Stiftloch
    Einfachspalt

Störung
    Verdichtungsstörung
    Verdünnungsstörung

Erdbeben
    Erdkern
    Erdkruste
    Erdmantel
    Epizentrum
    Hypozentrum
    Primärwelle
    Sekundärwelle
    seismische Welle
    Seismogramm

Fourier-Analyse
    Frequenzanteil
    Grundschwingung, Grundton
    Grundschwingung
    Grundton
    Oberschwingung, Oberton
    Oberschwingung, Oberton

phase (of a wave)	**Phase (einer Welle)**
in phase	in Phase
opposite in phase	gegenphasig
phase-shifted	phasenverschoben
sound wave	**Schallwelle**
sonar	Echolot
sonic	Schall-
sonic sender	Schallsender
tuning fork	Stimmgabel
standing wave	**stehende Welle**
antinode	Wellenbauch
Chladni patterns	Chladni'sche Klangfiguren
fixed end	festes Ende
free end	freies Ende
fundamental	Grundschwingung
fundamental oscillation	Grundschwingung
harmonic	Oberschwingung
Kundt's tube	Kundt'sches Rohr
natural oscillation	Eigenschwingung
nodal line	Knotenlinie
node	Wellenknoten
overtone	Oberschwingung
phase-dependent amplitude	phasenabhängige Amplitude
sawdust	Sägemehl
stationary wave	stehende Welle
superposition	**Superposition, Überlagerung**
to cancel each other out	sich gegenseitig auslöschen
constructive interference	konstruktive Interferenz
destructive interference	destruktive Interferenz
to interfere	interferieren
interference	Interferenz
path difference	Gangunterschied, Pfaddifferenz
phase difference	Phasendifferenz
types of waves	**Wellenarten**
circular wave	Kreiswelle
longitudinal wave	longitudinale Welle
pressure wave	Druckwelle
shear wave	Scherwelle
→ sound wave	Schallwelle
spheric wave	Kugelwelle
→ standing wave	stehende Welle
straight wave	gerade Welle

Waves / Wellen

transverse wave — transversale Welle
→ water wave — Wasserwelle

water wave — Wasserwelle
ripple tank — Wellenwanne
surface tension — Oberflächenspannung

wave — Welle
amplitude — Amplitude
angular frequency — Kreisfrequenz
angular speed — Winkelgeschwindigkeit
angular velocity — Vektor der Winkelgeschwindigkeit
celerity — Phasengeschwindigkeit
collimation — Kollimation
collimator — Kollimator
crest — Wellenberg
diffusion (of a wave) — Ausbreitung (einer Welle)
dispersion — Dispersion, Streuung
elongation — Auslenkung, Elongation
to excite (wave) — anregen (Welle)
exciter (of a wave) — Erreger (einer Welle)
frequency — Frequenz
harmonic wave — harmonische Welle
intensity — Intensität
oscillator velocity — Oszillatorengeschwindigkeit, Schnelle (Oszillatorengeschwindigkeit)

period — Schwingungsdauer
→ phase (of a wave) — Phase (einer Welle)
phase velocity — Phasengeschwindigkeit, Vektor der Phasengeschwindigkeit

point source — punktförmige Quelle
power of a wave — Leistung einer Welle
propagation (of a wave) — Ausbreitung (einer Welle)
propagation speed — Fortpflanzungsgeschwindigkeit
reflection — Reflexion
scattering — Streuung
trough — Wellental
wave crest — Wellenberg
wave equation — Wellengleichung
wave front — Wellenfront
wave number — Wellenzahl
wave trough — Wellental
wavelength — Wellenlänge
wavenumber — Wellenzahl

Acoustics / Akustik

Acoustics — Akustik

ear — Ohr
- anvil (ear) — Amboss (Ohr)
- cochlea — Gehörschnecke
- cochlear nerve — Hörnerv
- ear canal — Gehörgang
- ear muffs — Ohrenschützer
- ear ossicles — Gehörknöchelchen
- ear plugs — Ohrenstopfen, Ohrenstöpsel
- eardrum — Trommelfell (Ohr)
- Eustachian tube (ear) — eustachische Röhre (Ohr), Ohrtrompete (Ohr)
- hammer (ear) — Hammer (Ohr)
- hearing damage — Gehörschädigung
- hearing threshold — Hörschwelle
- inner ear — Innenohr
- middle ear — Mittelohr
- middle ear cavity — Paukenhöhle (Ohr)
- outer ear — Außenohr
- oval window (ear) — ovales Fenster (Ohr)
- pain limit — Schmerzgrenze
- semicircular canals (ear) — Bogengänge (Ohr)
- stirrup (ear) — Steigbügel (Ohr)

guitar — Gitarre
- bass guitar — Bassgitarre
- bass string — Basssaite
- bridge (guitar) — Steg (Gitarre)
- fret (guitar) — Bund (Gitarre), Bundstab (Gitarre)
- guitar string — Gitarrensaite
- linear density (of a guitar string) — Längendichte (einer Gitarrensaite)
- nut (guitar) — Sattel (Gitarre)
- tuning key (guitar) — Wirbel (Gitarre)

music — Musik
- acoustic colour — Klangfarbe
- chamber pitch — Kammerton
- fundamental — Grundton
- fundamental tone — Grundton
- harmonic — Oberton
- major scale — Dur-Tonleiter
- musical scale — Tonleiter
- note (C, D, E …) — Note (c, d, e …)
- octave — Oktave

Acoustics / Akustik

overtone	Oberton
pitch	Tonhöhe
semitone	Halbton
timbre	Klang
tone	Ton
tone pitch	Tonhöhe

sound — **Schall**

infrasonic	Infraschall-
infrasound	Infraschall
intensity	Intensität
loudness	Lautstärkepegel
microphone	Mikrophon
sonar	Echolot
sonic	Schall-
sonic sender	Schallsender
sound level	Schallpegel
sound pressure	Schalldruck
sound pressure level	Schalldruckpegel, Schallpegel
sound wave	Schallwelle
SPL (sound pressure level)	Schalldruckpegel, Schallpegel
ultrasonic	Ultraschall-
ultrasound	Ultraschall

sound generator — **Schallerzeuger, Tonerzeuger**

loudspeaker	Lautsprecher
piano string	Klaviersaite
pipe organ	Pfeifenorgel
punched disc	Lochscheibe
siren	Sirene
siren disc	Lochsirene
string	Saite
string instrument	Saiteninstrument, Streichinstrument
stringed instrument	Streichinstrument
tuning fork	Stimmgabel
vocal cord	Stimmband
wind instrument	Blasinstrument

Radioactivity

atom[10]
- atomic number
- atomic structure
- average mass number
- charge number
- electron number
- element
- element name
- emission
- to emit
- energy level
- excitation (atom, electron, nucleus)
- to excite (atom, electron, nucleus)
- excited (atom, electron, nucleus)
- excited state
- ground level
- ground state
- nomenclature
- photon
- photon energy
- structured

atomic nucleus
- atomic nuclei
- isotope
- mass number
- neutron
- neutron number
- nuclear force
- nuclei
- nucleon
- nucleus
- nuclide
- proton
- proton number
- table of nuclides

C-14 dating
- assimilation
- C-14 portion

Radioaktivität

Atom
- Ordnungszahl
- Atombau, Atomstruktur
- mittlere Massenzahl
- Ladungszahl
- Elektronenzahl
- Element
- Elementname
- Emission
- emittieren
- Energieniveau
- Anregung (Atom, Elektron, Atomkern)
- anregen (Atom, Elektron, Atomkern)
- angeregt (Atom, Elektron, Atomkern)
- angeregter Zustand
- Grundzustand
- Grundzustand
- Nomenklatur
- Photon
- Photonenenergie
- aufgebaut, strukturiert

Atomkern
- Atomkerne
- Isotop
- Massenzahl
- Neutron
- Neutronenzahl
- Kernkraft
- Atomkerne, Kerne
- Nukleon
- Atomkern, Kern
- Nuklid
- Proton
- Protonenzahl
- Nuklidtafel

C-14-Methode
- Assimilation
- C-14-Anteil

---

[10] see also chapter quantum physics

Radioactivity / Radioaktivität

fauna	Fauna
flora	Flora
radiocarbon dating	Radiokarbonmethode

cloud chamber — Nebelkammer

condensation nucleus	Kondensationskern
humidity	Luftfeuchtigkeit
ion	Ion
particle track	Teilchenspur
supersaturated	übersättigt
supersaturated vapour	übersättigter Dampf
supersaturated water vapour	übersättigter Wasserdampf

Geiger-Müller tube — Geiger-Müller-Zähler

electron avalanche	Elektronenlawine
electronic signal	elektronisches Signal
ionisation	Ionisation
primary electron	primäres Elektron
secondary electron	sekundäres Elektron

nuclear power plant — Atomkraftwerk, Kernkraftwerk

fissile	spaltbar
heavy water	schweres Wasser
nuclear fission	Kernspaltung
nuclear fusion	Kernfusion
worst case scenario	GAU, größter annehmbarer Unfall

radiation — radioaktive Strahlung, Strahlung

absorbed dose	Energiedosis
absorption	Absorption
alpha particle	Alphateilchen
background radiation	natürliche Radioaktivität
beta particle	Betateilchen
cosmic radiation	kosmische Strahlung
count rate	Zählrate
energy-rich	energiereich
equivalent dose	Äquivalentdosis
gamma particle	Gammateilchen
gray	Gray
natural background radiation	natürliche Radioaktivität
nuclear radiation	radioaktive Strahlung
primary particle	Primärteilchen
radiation shielding	Abschirmung radioaktiver Strahlung
secondary particle	Sekundärteilchen
sievert	Sievert
terrestrial radiation	Bodenstrahlung

radioactive decay
- alpha decay
- beta decay
- curie
- daughter nucleus
- decay
- to decay
- decay chain
- decay constant
- decay mode
- decomposition
- disintegration
- equation
- gamma decay
- main decay mode
- parent nucleus
- → radiation
- unstable

radioactivity
- activity
- alpha radioactivity
- artificial radiation source
- becquerel
- beta radioactivity
- compound
- gamma radioactivity
- half life
- initial activity
- → radiation
- radioactive
- → radioactive decay
- radioactive isotope

rules
- activity law
- radioactive decay law

uranium-lead method
- magma
- ore
- pitchblende
- uranium ore
- uranium rock

radioaktiver Zerfall
- Alphazerfall
- Betazerfall
- Curie
- Tochterkern
- Zerfall
- zerfallen
- Zerfallsreihe
- Zerfallskonstante
- Zerfallsmodus
- radioaktiver Zerfall, Zerfall
- radioaktiver Zerfall, Zerfall
- Reaktionsgleichung
- Gammazerfall
- Hauptzerfallsmodus
- Mutterkern
- radioaktive Strahlung
- instabil

Radioaktivität
- Aktivität
- Alpharadioaktivität
- künstliche Quelle radioaktiver Strahlung
- Becquerel
- Betaradioaktivität
- Präparat
- Gammaradioaktivität
- Halbwertszeit
- anfängliche Aktivität
- radioaktive Strahlung
- radioaktiv
- radioaktiver Zerfall
- radioaktives Isotop

Regeln
- Aktivitätsgesetz
- radioaktives Zerfallsgesetz

Uran-Blei-Methode
- Magma
- Erz
- Pechblende
- Uranerz
- Urangestein

Radioactivity / Radioaktivität

X-ray — Röntgenstrahl
- to have an X-ray — sich röntgen lassen
- to take an X-ray of someone — jemanden röntgen
- to X-ray — röntgen
- X-ray photon — Röntgenphoton
- X-ray quantum — Röntgenquant

Optics / Optik

absorption — Absorption
- absorbance — Absorptionsgrad
- optical density — optische Dichte (Extinktion)
- optical depth — optische Dicke, optische Tiefe
- optical thickness — optische Dicke, optische Tiefe
- transmittance — Transmissionsgrad

colour — Farbe
- additive mixture — additive Mischung
- body colours — Körperfarben
- colour addition — Farbaddition
- colour filter — Farbfilter
- colour subtraction — Farbsubtraktion
- coloured light — farbiges Licht
- two colours form a third — zwei Farben ergeben eine dritte
- two colours produce a third — zwei Farben ergeben eine dritte
- two colours yield a third — zwei Farben ergeben eine dritte
- complementary colours — Komplementärfarben
- cyan — cyan
- ink cartridge — Druckerpatrone, Tintenpatrone (Drucker)
- magenta — magenta
- to mix coloured light — farbiges Licht mischen
- paintbox — Malkasten
- primary colour — Grundfarbe
- printing method — Druckverfahren
- secondary colour — Mischfarbe
- slide — Dia
- → spectrum — Spektrum
- subtractive mixture — subtraktive Mischung
- toner cartridge — Tonerkartusche, Tonerpatrone
- turquoise — türkis
- yellow — gelb, yellow

Optics / Optik

diffraction[11] — Beugung

emission — Emission
- excitance (power per surface area) — Ausstrahlung (Leistung pro Oberfläche)
- → light source — Lichtquelle
- radiance (power per area per solid angle) — Strahldichte (Leistung pro Fläche pro Raumwinkel)
- radiation — Strahlung

eye — Auge
- accommodation — Akkommodation
- ciliary muscle — Ziliarmuskel
- closest focusing distance — minimale Sehweite, minimale Sichtdistanz
- cones (eye) — Zapfen (Auge)
- cornea (eye) — Hornhaut (Auge)
- eye lens — Augenlinse
- eyeball — Augapfel
- hyperopic — weitsichtig
- iris — Iris
- long-sighted — weitsichtig
- long-sightedness — Weitsichtigkeit
- myopic — kurzsichtig
- retina — Netzhaut, Retina
- rods (eye) — Stäbchen (Auge)
- sensory cells — Sinneszellen
- short-sighted — kurzsichtig
- short-sightedness — Kurzsichtigkeit

image formation — Abbildung
- to blur — verschwimmen
- blurred — unscharf, verschwommen
- centre ray — Zentralstrahl
- enlarged image — vergrößertes Bild
- focal plane — Brennebene
- focal point — Brennpunkt
- focal ray — Brennstrahl
- image — Bild
- image distance — Bildweite
- image formation equation — Abbildungsgleichung
- image orientation — Bildorientierung
- image size — Bildgröße
- intermediate image — Zwischenbild
- inverted image — umgekehrtes Bild

---

[11] see chapter waves

Optics / Optik

    magnification — Vergrößerung
    object — Gegenstand
    object distance — Gegenstandsweite
    object size — Gegenstandsgröße
    parallel ray — Parallelstrahl
    real image — reelles Bild
    same size image — Bild gleicher Größe
    screen — Schirm
    sharpness — Schärfe
    to soften (sharpness) — sich abschwächen (Schärfe)
    upright image — aufrechtes Bild
    virtual focal point — virtueller Brennpunkt
    virtual image — virtuelles Bild

incidence — Einfall (von Licht)
    illuminance — Beleuchtungsstärke
    illumination — Beleuchtung
    incidence angle — Einfallswinkel
    incident ray — einfallender Strahl
    irradiance (power per area) — Einstrahlung (Leistung pro Fläche)
    normal — Normale

lens — Linse
    concave lens — Konkavlinse
    converging lens — Sammellinse
    convex lens — Konvexlinse
    dioptre — Dioptrie
    dioptres — Dioptrien
    diverging lens — Zerstreuungslinse
    focal length — Brennweite
    focus — Brennpunkt
    lens combination — Linsenkombination
    lens distance — Linsenabstand
    lens equation — Linsengleichung
    lens system — Linsensystem
    optical power — Brechkraft
    principal focus — Brennpunkt

light propagation — Lichtausbreitung
    convergent — konvergent
    divergent — divergent
    light ray — Lichtstrahl
    light rays coincide — Lichtstrahlen treffen aufeinander, Lichtstrahlen treffen sich

    parallel [adj.] [adv.] — parallel

principle of the reversibility of the light path	Prinzip der Umkehrbarkeit des Lichtwegs
to proceed straight	geradeaus verlaufen
reversibility of the light path	Umkehrbarkeit des Lichtwegs
speed of light in a medium	Lichtgeschwindigkeit in einem Medium
speed of light in vacuum	Lichtgeschwindigkeit im Vakuum

light source / Lichtquelle

bright	hell
brightness	Helligkeit
fluorescent lamp	Leuchtstoffröhre
halogen lamp	Halogenlampe
hydrogen lamp	Wasserstofflampe
light bulb	Glühbirne
a light bulb shines	eine Glühbirne leuchtet
luminescent layer (of a fluorescent lamp)	Leuchtschicht (einer Leuchtstoffröhre)
luminous object	Selbstleuchter
neon lamp	Neonlampe
non-luminous object	Fremdleuchter
phosphorescence	Nachleuchten, Phosphoreszenz
to shine (e. g. a light bulb)	leuchten (z. B. eine Glühbirne)
sodium lamp	Natriumdampflampe

mirror / Spiegel

concave mirror	Hohlspiegel
concave mirror equation	Hohlspiegelgleichung
convex mirror	Wölbspiegel
mirrored	verspiegelt
parabolic dish	Parabolspiegel
parabolic mirror	Parabolspiegel

optical bench / optische Bank

to adjust	justieren
to align	ausrichten
aperture	Apertur, Blende
arrow aperture	Pfeilblende
candle holder	Kerzenhalter
centimetre grid	Zentimeterraster
centric	zentrisch
diffuser	Diffusor
holder (e. g. for a slit aperture)	Halter (z. B. für eine Spaltblende)
multi-slit aperture	Mehrfachspaltblende
optical axis	optische Achse
principal axis	optische Achse
to readjust	neu justieren

Optics / Optik

to slide	gleiten
slider (on an optical bench)	Reiter (auf einer optischen Bank)
slit aperture	Spaltblende

optical instrument — optisches Instrument

binoculars	Fernglas
eyepiece	Okular
field of view	Sehwinkel
Galilean telescope	Galileisches Fernrohr
Galileo's telescope	Galileisches Fernrohr
glasses	Brille
Kepler telescope	Kepler'sches Fernrohr
magnifying glass	Lupe
microscope	Mikroskop
objective	Objektiv
pinhole camera	Lochkamera
reticle	Fadenkreuz
telescope	Fernrohr

radiative transfer — Strahlungsdurchgang

→ absorption	Absorption
→ emission	Emission
scattering	Streuung
visibility	Sichtweite

reflection — Reflexion, Spiegelung

boundary	Grenzfläche
dark	dunkel
diffuse reflection	diffuse Reflexion
→ incidence	Einfall (von Licht)
irregular reflection	unregelmäßige Reflexion
law of reflection	Reflexionsgesetz
to reflect	reflektieren
reflectance	Reflektanz, Reflexionsgrad
reflected ray	reflektierter Strahl
reflection angle	Reflexionswinkel
reflectivity	Reflektanz
regular reflection	regelmäßige Reflexion
rough surface	raue Oberfläche
shiny	glänzend
smooth surface	glatte Oberfläche
specular reflection	regelmäßige Reflexion

refraction — Brechung

cedar oil	Zedernöl
double refraction	Doppelbrechung

    law of refraction   Brechungsgesetz
    refraction angle   Brechungswinkel
    refractive index   Brechzahl, optische Dichte (Brechzahl)
    transition   Übergang

shadow   Schatten

    to cast a shadow   einen Schatten werfen
    full solar eclipse   totale Sonnenfinsternis
    partial solar eclipse   partielle Sonnenfinsternis
    penumbra   Halbschatten
    shadow formation equation   Schattenbildungsgleichung
    shadow space   Schattenraum
    umbra   Kernschatten

spectrum   Spektrum

    dispersion (chromatic dispersion of light)   Dispersion (chromatische Dispersion von Licht)
    Fraunhofer line   Fraunhoferlinie
    line spectrum   Linienspektrum
    prism   Prisma
    rainbow   Regenbogen
    secondary rainbow   sekundärer Regenbogen
    spectra   Spektren
    spectral colours   Spektralfarben
    spectroscope   Spektroskop
    spectrum of visible light   Spektrum des sichtbaren Lichts

total reflection   Totalreflexion

    critical angle of total reflection   kritischer Winkel der Totalreflexion, Winkel der Totalreflexion
    total internal reflection   Totalreflexion

Relativity

beta factor	Beta-Faktor
blueshift	Blauverschiebung
carrier medium	Trägermedium
classical relativity	klassische Relativität
concurrency	Gleichzeitigkeit
Galilean transformation	Galilei-Transformation
general theory of relativity	allgemeine Relativitätstheorie
inertia	Trägheit
inertial system	Inertialsystem
inverse Galilean transformation	inverse Galilei-Transformation
length contraction	Längenkontraktion
Lorentz factor	Lorentz-Faktor
mass-energy equivalence	Masse-Energie-Äquivalenz
postulates of relativity	Postulate der Relativitätstheorie
principle of relativity	Relativitätsprinzip
proper length	Eigenlänge
proper time	Eigenzeit
redshift	Rotverschiebung
relativistic energy-momentum relation	relativistische Energie-Impuls-Beziehung
relativistic kinetic energy	relativistische kinetische Energie
relativistic mass	relativistische Masse
relativistic momentum	relativistischer Impuls
relativity	Relativität
relativity principle	Relativitätsprinzip
to rest	ruhen
rest energy	Ruheenergie
rest frame	Ruhesystem
rest mass	Ruhemasse
special theory of relativity	spezielle Relativitätstheorie
system	System
theory of relativity	Relativitätstheorie
time dilation	Zeitdilatation

Quantum Physics

atom
- allowed orbit
- atomic number
- atomic shell
- atomic structure
- average mass number
- bound electron
- charge number
- electron
- electron excitation
- electron number
- element
- element name
- emission
- to emit
- energy level
- excitation (atom, electron, nucleus)
- to excite (atom, electron, nucleus)
- excited (atom, electron, nucleus)
- excited state
- ground level
- ground state
- ionisation
- nomenclature
- photon
- photon energy
- quantum leap
- structured

rules
- de Broglie relation
- Heisenberg uncertainty principle
- Schrödinger equation

miscellaneous
- electron volt
- Planck constant
- quantum
- quantum mechanics
- quantum number
- spin

Quantenphysik

Atom
- erlaubte Bahn
- Ordnungszahl
- Atomhülle, Atomschale
- Atombau, Atomstruktur
- mittlere Massenzahl
- gebundenes Elektron
- Ladungszahl
- Elektron
- Elektronenanregung
- Elektronenzahl
- Element
- Elementname
- Emission
- emittieren
- Energieniveau
- Anregung (Atom, Elektron, Atomkern)
- anregen (Atom, Elektron, Atomkern)
- angeregt (Atom, Elektron, Atomkern)
- angeregter Zustand
- Grundzustand
- Grundzustand
- Ionisation
- Nomenklatur
- Photon
- Photonenenergie
- Quantensprung
- aufgebaut, strukturiert

Regeln
- de Broglie-Beziehung
- Heisenberg'sche Unschärferelation
- Schrödingergleichung

Verschiedenes
- Elektronenvolt
- Planck'sches Wirkungsquantum
- Quant
- Quantenmechanik
- Quantenzahl
- Spin

Physical Constants
Physikalische Konstanten

atomic mass unit	atomare Masseneinheit
Avogadro's number	Avogadro-Zahl
Boltzmann constant	Boltzmann-Konstante
charge-to-mass ratio for the electron	Ladungs-Massenverhältnis des Elektrons
elementary charge	Elementarladung
gas constant	allgemeine Gaskonstante
gravitational constant	Gravitationskonstante
permeability of vacuum	magnetische Feldkonstante
permittivity of vacuum	elektrische Feldkonstante
Planck constant	Planck'sches Wirkungsquantum
standard acceleration due to gravity	Normwert der Fallbeschleunigung
standard pressure	Normdruck
standard temperature	Normtemperatur
Stefan-Boltzmann constant	Stefan-Boltzmann-Konstante

SI-base Quantities
SI-Basisgrößen

SI-base quantity — SI-Basisgröße
- length — Länge
- time — Zeit
- mass — Masse
- temperature — Temperatur
- amount of substance — Stoffmenge
- electric current — elektrische Stromstärke
- luminous intensity — Luminosität

Other physical Quantities
Andere physikalische Größen

absorbed dose	Energiedosis
acceleration	Beschleunigung
activity	Aktivität
angular speed	Winkelgeschwindigkeit
area	Fläche
capacitance	Kapazität (eines Kondensators)
density	Dichte
dynamic viscosity	dynamische Viskosität

electric charge	elektrische Ladung
energy	Energie
equivalent dose	Äquivalentdosis, Äquivalenzdosis
flow	Fluss
force	Kraft
frequency	Frequenz
impulse	Kraftstoß
inductance	Induktivität
intensity	Intensität
magnetic field	Magnetfeld
magnetic flux	magnetischer Fluss
magnetic flux density	magnetische Flussdichte
momentum	Impuls
power	Leistung
pressure	Druck
radioactivity	Radioaktivität
resistance	Widerstand (physikalische Größe)
speed	Geschwindigkeit
torque	Drehmoment
voltage	Spannung
volume	Volumen

SI-base Units / SI-Grundeinheiten

SI-base unit	SI-Grundeinheit
metre	Meter
second	Sekunde
kilogram	Kilogramm
kelvin	Kelvin
mole	Mol
ampere	Ampere
candela	Candela

Derived SI-Units / Abgeleitete SI-Einheiten

becquerel	Becquerel
coulomb	Coulomb
cubic metre	Kubikmeter
cubic metre per second	Kubikmeter pro Sekunde
farad	Farad
gray	Gray
henry	Henry

Derived SI-Units / Abgeleitete SI-Einheiten

hertz	Hertz
joule	Joule
kilogram per cubic metre	Kilogramm pro Kubikmeter
metre per second	Meter pro Sekunde
metre per second squared	Meter pro Quadratsekunde
newton	Newton
newton metre	Newtonmeter
ohm	Ohm
pascal	Pascal
pascal second	Pascalsekunde
sievert	Sievert
square metre	Quadratmeter
tesla	Tesla
volt	Volt
watt	Watt
weber	Weber

Non-SI-Units / Nicht-SI-Einheiten

acre	Morgen
angstrom	Angström
are	Are
astronomical unit	astronomische Einheit
atmosphere	Atmosphäre
atomic unit of charge	atomare Ladungseinheit
bar	Bar
barrel	Barrel
calorie	Kalorie
carat	Karat
curie	Curie
day	Tag
degree Celsius	Grad Celsius
degree Fahrenheit	Grad Fahrenheit
electron volt	Elektronenvolt
foot	Fuß
gallon	Gallone
grain	Gran
hectare	Hektar
horsepower	Pferdestärke
hour	Stunde
inch	Zoll
Julian year	Julianisches Jahr
kilometre per hour	Kilometer pro Stunde
kilowatt-hour	Kilowattstunde

knot	Knoten
light-year	Lichtjahr
litre	Liter
metric tonne	Tonne
mile	Meile
millimetre of mercury	Millimeter Quecksilbersäule
minute	Minute
nautical mile	nautische Meile
ounce	Unze
pint	Pint
pound	Pfund
pound-force	Pound-force (US-amerikanische Krafteinheit)
pound-force per square inch	psi (US-amerikanische Druckeinheit)
psi	psi (US-amerikanische Druckeinheit)
quart	Quart
speed of light in vacuum	Lichtgeschwindigkeit im Vakuum
tablespoon	Esslöffel
teaspoon	Teelöffel
troy ounce	Feinunze
troy pound	Apotheker-Pfund
yard	Yard

Technical Terms / Technische Begriffe

Technical Terms / Technische Begriffe

aviation / Luftfahrt

English	Deutsch
aeroplane [B. E.]	Flugzeug
airplane [A. E.]	Flugzeug
catapult seat	Schleudersitz
ejection seat	Schleudersitz
glider	Segelflugzeug
kite	Drachen
parachute	Fallschirm
runway	Landebahn, Startbahn
sailplane	Segelflugzeug
wing (aeroplane)	Tragfläche (Flugzeug)

engineering / Maschinenbau

English	Deutsch
bolt	Bolzen (mit Gewinde), Metallschraube
component	Bauteil
craftsman	Handwerker
extension	Ausdehnung
exterior	Äußeres
function principle	Funktionsprinzip
functionality	Funktionsweise
genuine part	Originalteil
interior	Inneres
iron girder	Eisenträger
levelling screw	Stellschraube (zur Höheneinstellung)
makeup	Aufbau
material control	Materialprüfung
nut	Mutter
part	Bauteil, Teil
pin	Bolzen (ohne Gewinde)
robot	Roboter
screw	Holzschraube
spare part	Ersatzteil
technique	Technik
template	Muster
thread	Gewinde
→ tool	Werkzeug

equipment / Gerät

English	Deutsch
capsule	Kapsel
carpet cleaner [A. E.]	Staubsauger
conveyor belt	Förderband
cupola	Kuppel
elevator [A. E.]	Aufzug

Themenspezifische Wortverzeichnisse

elevator cab [A. E.]	Aufzugkabine
elevator cabin [A. E.]	Aufzugkabine
fire hose	Feuerwehrschlauch
flywheel	Schwungrad
gas cylinder	Gasflasche
handlebars	Lenkstange (Fahrrad)
Hoover [B. E.]	Staubsauger
ice skate	Schlittschuh
kettle	Wasserkessel
knitting needle	Stricknadel
lamp mounting	Lampenaufhängung
lid	Deckel
lift [B. E.]	Aufzug
lift car [B. E.]	Aufzugkabine
light fixture	Lampenaufhängung
net	Netz
paperclip	Büroklammer
photographic plate	Fotoplatte
ping-pong ball	Tischtennisball
rope	Seil
smoke detector	Rauchmelder
steel cable	Drahtseil, Stahlseil
steel rope	Drahtseil, Stahlseil
stirrup	Steigbügel
straw	Strohhalm
tap	Wasserhahn
trampoline	Trampolin
vacuum cleaner	Staubsauger
wax ball	Wachskugel
wrist watch	Armbanduhr

liquids / Flüssigkeiten

cedar oil	Zedernholzöl
crude oil	Rohöl
diesel	Diesel
gas oil	Zweitaktöl
glycerin	Glycerin, Glycerol, Glyzerin
glycerine	Glycerin, Glycerol, Glyzerin
glycerol	Glycerin, Glycerol, Glyzerin
kerosene	Kerosin
linseed oil	Leinsamenöl
olive oil	Olivenöl
paraffin	Paraffin
petroleum	Petroleum
rape seed oil	Rapssamenöl

Technical Terms / Technische Begriffe

silicone oil	Silikonöl
soy been oil	Sojabohnenöl
soya been oil	Sojabohnenöl

materials / Materialien

amber	Bernstein
asphalt	Asphalt
bone	Knochen
brass	Messing
carbon	Kohle
carbon steel	Karbonstahl
cast iron	Grauguss
cast steal	Gusseisen
ceramics	Keramik
clay	Ton
clay brick	Ziegelstein
concrete	Beton
cotton cloth	Baumwolltuch
cotton wool	Watte
diamond	Diamant
fibreglass	Glasfaser
flint	Feuerstein
flint glass	Flintglas
glass	Glas
granite	Granit
gravel	Kies, Schotter
grit	Splitt
hard rubber	Hartgummi
invar	Invar
leather	Leder
marble	Marmor
mica	Glimmer
nylon	Nylon
opal	Opal
paraffin	Paraffin
plastic	Plastik
Plexiglas	Plexiglas
plexiglass	Plexiglas
polystyrene	Styropor
porcelain	Porzellan
quartz	Quarz
quartz glass	Quarzglas
sand	Sand
sapphire	Saphir
steel	Stahl

Themenspezifische Wortverzeichnisse

styrofoam	Styropor
tar	Teer
to tar	teeren
tarmac	Asphalt
to tarmac	asphaltieren
Teflon	Teflon
textiles	Textilien
wax ball	Wachskugel
window glass	Fensterglas
→ wood	Holz

property / Eigenschaft

arched	gewölbt
brittle	spröde
curved	gewölbt
pointed	spitz
polished	poliert
rigid	steif
sharp	scharf, spitz
spiky	spitz

railway [B. E.] / Eisenbahn

boxcar	Güterwagen
buffer	Puffer
to derail	entgleisen
locomotion	Fortbewegung
locomotive	Lokomotive
rail	Schiene
railroad [A. E.]	Eisenbahn
railway line	Bahnlinie, Eisenbahnstrecke
railway track	Gleisanlage, Schiene
track	Bahngleis
train carriage	Eisenbahnwagen
train wagon	Eisenbahnwagon

shipping / Schifffahrt

anchor	Anker
to anchor	verankern
cargo	Fracht
cargo ship	Frachter, Frachtschiff
helm	Steuerruder
mast	Mast
oar	Ruder
payload	Nutzlast
tug	Schleppschiff
tugboat	Schleppschiff

Technical Terms / Technische Begriffe

technical process — technischer Vorgang

English	German
to age	altern
ageing	Altern
aging	Altern
to aluminize	aluminieren
aluminized	aluminiert
to assemble	zusammenbauen
to blast	sprengen
to blow	sprengen
to blow up	sprengen
to break	zerbrechen
to bust	sprengen
to coil itself up	sich aufwickeln
to coil up something	etwas aufwickeln
to collide	zusammenstoßen
collision	Zusammenstoß
to crack	zerbrechen
to cut a thread	ein Gewinde schneiden
lacquer	Lack
to manoeuver	manövrieren
to polish	polieren
to rebound	abprallen
to varnish	lackieren

tool — Werkzeug

English	German
anvil (tool)	Amboss (Werkzeug)
brush	Bürste
to drill	bohren
drill	Bohrer, Bohrmaschine
to grind	schleifen
grinding stone	Schleifstein
grindstone	Schleifstein
hammer (tool)	Hammer (Werkzeug)
hammer handle	Hammerstiel
hammer head	Hammerkopf
level	Wasserwaage
Phillips screwdriver	Kreuzschraubendreher
pincers	Beißzange
pliers	Zange
saw	Säge
to saw	sägen
sawdust	Sägemehl
scissors	Schere
screwdriver	Schraubendreher
to solder	löten

Themenspezifische Wortverzeichnisse

soldering iron	Lötkolben
syringe	Spritze
vice	Schraubstock
to weld	schweißen
welder	Schweißgerät
winch	Seilwinde, Winde
to wind up something	etwas aufwickeln
wrench	Schraubenschlüssel

vehicle / Fahrzeug

bumper	Stoßstange
caterpillar	Raupe
crane	Kran
driven axle	angetriebene Achse
excavator	Bagger
excavator shovel	Baggerschaufel
fuel	Treibstoff
fuel gauge	Tankuhr
motorcycle	Motorrad
petrol gauge	Tankuhr
rim	Felge
roller coaster	Achterbahn
wheel rim	Felge

weapon / Waffe

blow pipe	Blasrohr
blowgun	Blasrohr
bow and arrow	Pfeil und Bogen
bullet	Kugel (Geschoss)
long-range missile	Langstreckenrakete
missile	Rakete (Waffe)
nuclear warhead	Atomsprengkopf
target	Ziel, Zielscheibe

wood / Holz

against the grain	quer zur Faser
along the grain	parallel zur Faser
beech wood	Buchenholz
birch wood	Birkenholz
cork	Kork
oak wood	Eichenholz
planed	gehobelt
spruce wood	Fichtenholz

Chemistry

atom[12]
    atomic nuclei
    atomic nucleus
    atomic number
    atomic shell
    atomic structure
    average mass number
    bound electron
    charge number
    electron
    electron number
    energy level
    neutron
    neutron number
    nomenclature
    nuclei
    nucleon
    nucleus
    proton
    proton number
    structured

chemical bond
    atomic lattice
    binding forces
    bond
    bound electron
    carbon-carbon bond
    to combine with
    combined
    combined with the lattice
    covalent bond
    crystal
    crystal lattice
    double bond
    equation
    ionic bond
    lattice
    metallic bond
    pi bond

Chemie

Atom
    Atomkerne
    Atomkern
    Ordnungszahl
    Atomhülle, Atomschale
    Atombau, Atomstruktur
    mittlere Massenzahl
    gebundenes Elektron
    Ladungszahl
    Elektron
    Elektronenzahl
    Energieniveau
    Neutron
    Neutronenzahl
    Nomenklatur
    Atomkerne, Kerne
    Nukleon
    Atomkern, Kern
    Proton
    Protonenzahl
    aufgebaut, strukturiert

chemische Bindung
    Atomgitter
    Bindungskräfte
    Bindung
    gebundenes Elektron
    C-C-Bindung
    sich verbinden mit
    gebunden
    gebunden an das Gitter
    kovalente Bindung
    Kristall
    Kristallgitter
    Doppelbindung
    Reaktionsgleichung
    Ionenbindung
    Gitter (eines Festkörpers), Kristallgitter
    Metallbindung
    Pi-Bindung

---

[12] see also chapter quantum physics

polar	polar
polarisation	Polarisation
salt	Salz
salt crystal	Salzkristall
sigma bond	Sigma-Bindung
table salt crystal	Kochsalzkristall
uncombined	ungebunden

chemical reaction / chemische Reaktion

oxidation	Oxidation
to oxidise	oxidieren
oxidised	oxidiert
to reduce	reduzieren
reduced	reduziert
reduction	Reduktion

element / Element

chemical symbol	Elementsymbol
element name	Elementname
→ periodic system of the elements	Periodensystem

ionisation / Ionisation

anion	Anion
cation	Kation
ion	Ion

periodic system of the elements / Periodensystem

hydrogen [H, Z=1]	Wasserstoff
helium [He, Z=2]	Helium
lithium [Li, Z=3]	Lithium
beryllium [Be, Z=4]	Beryllium
boron [B, Z=5]	Bor
carbon [C, Z=6]	Kohlenstoff
nitrogen [N, Z=7]	Stickstoff
oxygen [O, Z=8]	Sauerstoff
fluorine [F, Z=9]	Fluor
neon [Ne, Z=10]	Neon
sodium [Na, Z=11]	Natrium
magnesium [Mg, Z=12]	Magnesium
aluminium [Al, Z=13]	Aluminium
silicon [Si, Z=14]	Silizium
phosphorus [P, Z=15]	Phosphor
sulphur [S, Z=16]	Schwefel
chlorine [Cl, Z=17]	Chlor
argon [Ar, Z=18]	Argon
potassium [K, Z=19]	Kalium

Chemistry / Chemie

calcium [Ca, Z=20]	Calcium
scandium [Sc, Z=21]	Scandium
titanium [Ti, Z=22]	Titan
vanadium [V, Z=23]	Vanadium
chromium [Cr, Z=24]	Chrom
manganese [Mn, Z=25]	Mangan
iron [Fe, Z=26]	Eisen
cobalt [Co, Z=27]	Kobalt
nickel [Ni, Z=28]	Nickel
copper [Cu, Z=29]	Kupfer
zinc [Zn, Z=30]	Zink
gallium [Ga, Z=31]	Gallium
germanium [Ge, Z=32]	Germanium
arsenic [As, Z=33]	Arsen
selenium [Se, Z=34]	Selen
bromine [Br, Z=35]	Brom
krypton [Kr, Z=36]	Krypton
rubidium [Rb, Z=37]	Rubidium
strontium [Sr, Z=38]	Strontium
yttrium [Y, Z=39]	Yttrium
zirconium [Zr, Z=40]	Zirkon
niobium [Nb, Z=41]	Niob
molybdenum [Mo, Z=42]	Molybdän
technetium [Tc, Z=43]	Technetium
ruthenium [Ru, Z=44]	Ruthenium
rhodium [Rh, Z=45]	Rhodium
palladium [Pd, Z=46]	Palladium
silver [Ag, Z=47]	Silber
cadmium [Cd, Z=48]	Kadmium
indium [In, Z=49]	Indium
tin [Sn, Z=50]	Zinn
antimony [Sb, Z=51]	Antimon
tellurium [Te, Z=52]	Tellur
iodine [I, Z=53]	Iod
xenon [Xe, Z=54]	Xenon
caesium [Cs, Z=55]	Cäsium
barium [Ba, Z=56]	Barium
lanthanum [La, Z=57]	Lanthan
cerium [Ce, Z=58]	Cer
praseodymium [Pr, Z=59]	Praseodym
neodymium [Nd, Z=60]	Neodym
promethium [Pm, Z=61]	Promethium
samarium [Sm, Z=62]	Samarium
europium [Eu, Z=63]	Europium
gadolinium [Gd, Z=64]	Gadolinium

terbium [Tb, Z=65]	Terbium
dysprosium [Dy, Z=66]	Dysprosium
holmium [Ho, Z=67]	Holmium
erbium [Er, Z=68]	Erbium
thulium [Tm, Z=69]	Thulium
ytterbium [Yb, Z=70]	Ytterbium
lutetium [Lu, Z=71]	Lutetium
hafnium [Hf, Z=72]	Hafnium
tantalum [Ta, Z=73]	Tantal
tungsten [W, Z=74]	Wolfram
rhenium [Re, Z=75]	Rhenium
osmium [Os, Z=76]	Osmium
iridium [Ir, Z=77]	Iridium
platinum [Pt, Z=78]	Platin
gold [Au, Z=79]	Gold
mercury [Hg, Z=80]	Quecksilber
thallium [Tl, Z=81]	Thallium
lead [Pb, Z=82]	Blei
bismuth [Bi, Z=83]	Bismut
polonium [Po, Z=84]	Polonium
astatine [At, Z=85]	Astat
radon [Rn, Z=86]	Radon
francium [Fr, Z=87]	Francium
radium [Ra, Z=88]	Radium
actinium [Ac, Z=89]	Actinium
thorium [Th, Z=90]	Thorium
protactinium [Pa, Z=91]	Protactinium
uranium [U, Z=92]	Uran
neptunium [Np, Z=93]	Neptunium
plutonium [Pu, Z=94]	Plutonium
americium [Am, Z=95]	Americium
curium [Cm, Z=96]	Curium
berkelium [Bk, Z=97]	Berkelium
californium [Cf, Z=98]	Californium
einsteinium [Es, Z=99]	Einsteinium
fermium [Fm, Z=100]	Fermium
mendelevium [Md, Z=101]	Mendelevium
nobelium [No, Z=102]	Nobelium
lawrencium [Lr, Z=103]	Lawrencium
rutherfordium [Rf, Z=104]	Rutherfordium
dubnium [Db, Z=105]	Dubnium
seaborgium [Sg, Z=106]	Seaborgium
bohrium [Bh, Z=107]	Bohrium
hassium [Hs, Z=108]	Hassium
meitnerium [Mt, Z=109]	Meitnerium

Chemistry / Chemie

substance	Substanz
acetone	Aceton
acetylene	Acetylen
air	Luft
ammonia	Ammoniak
benzene	Benzol
butane	Butan
carbon dioxide	Kohlendioxid
carbon monoxide	Kohlenmonoxid
carbon tetrachloride	Tetrachlorkohlenstoff
diethyl ether	Diethylether
dry ice	Trockeneis
ethane	Ethan
ethanol	Ethanol
freon	Freon
gas [A. E.]	Benzin
gasoline [A. E.]	Benzin
heptane	Heptan
hexane	Hexan
methane	Methan
methanol	Methanol
natural gas	Erdgas
nonane	Nonan
octane	Octan
paraffin oil	Paraffinöl
pentane	Pentan
petrol [B. E.]	Benzin
propane	Propan
sodium chloride	Natriumchlorid
sulfur dioxide	Schwefeldioxid
sulfuric acid	Schwefelsäure
table salt	Kochsalz
toluol	Toluol

Mathematics

algebra
- to deduce
- deduction
- → difference
- → equation
- to factorize
- → fraction
- → logarithm
- to map
- mapping
- → number
- → power
- → product
- → proportionality
- → quotient
- → real number
- → root
- rule of three
- set
- subset of
- → sum
- → vector

analysis
- → calculus
- → diagram
- function value
- polynomial
- real function
- → straight line

angle
- circular measure
- degree
- to enclose an angle with
- to encompass an angle with
- radian
- radian measure
- solid angle

calculus
- derivation
- to derive

Mathematik

Algebra
- ableiten (logisch), herleiten
- Herleitung
- Differenz
- Gleichung
- ausklammern, faktorisieren
- Bruch
- Logarithmus
- abbilden
- Abbildung
- Zahl
- Potenz
- Produkt
- Proportionalität
- Produkt
- reelle Zahl
- Wurzel
- Dreisatz
- Menge
- Teilmenge von
- Summe
- Vektor

Analysis
- Differenzial- und Integralrechnung
- Diagramm
- Funktionswert
- Polynom
- reelle Funktion
- Gerade

Winkel
- Bogenmaß
- Grad
- einen Winkel einschließen mit
- einen Winkel einschließen mit
- Radian
- Bogenmaß
- Raumwinkel

Differenzial- und Integralrechnung
- Ableitung
- ableiten, differenzieren

differential calculus	Differenzialrechnung
Fourier analysis	Fourier-Analyse
integral	Integral
integral calculus	Integralrechnung
to integrate	integrieren
integration	Integration

diagram — Diagramm

x-axis	x-Achse
y-axis	y-Achse
axis	Achse
chart	Diagramm
coordinate system	Koordinatensystem
graph	Diagramm, Graph, Kurve
hyperbola	Hyperbel
interval	Intervall
origin	Nullpunkt
parabola	Parabel
pie chart	Kreisdiagramm, Kuchendiagramm
to plot	graphisch darstellen
plot [A. E.]	Diagramm
sine curve	Sinuskurve
sinusoid	Sinusfunktion, Sinuskurve
sinusoidal	sinusförmig
zero point	Nullpunkt

difference — Differenz

minuend	Minuend
minus	minus
a minus b	a minus b
to subtract	subtrahieren
subtraction	Subtraktion
subtrahend	Subtrahend

equation — Gleichung

to equal	gleich sein
a minus b equals c	a minus b ist gleich c

fraction — Bruch

denominator	Nenner
a divided by b	a geteilt durch b
fraction bar	Bruchstrich
numerator	Zähler
a over b	a geteilt durch b
nth part	n-ter Teil
reciprocal	Kehrwert

geometric object
- circle
- cube
- cuboid
- cylinder
- full circle
- parallelogram
- rectangle
- semicircle
- sphere
- spiral
- square

geometry
- → angle
- arc length
- area
- circumference (especially of a circle)
- compass
- compasses
- to displace
- displaced
- displacement
- edge length
- → geometric object
- intersection
- length of arc
- pair of compasses [rare]
- parallel [adj.] [adv.]
- perimeter
- perpendicular
- perpendicularly
- radius
- semicircular
- to shift
- shifted
- spacial
- surface area
- tangential
- tangentially
- → trigonometry
- unit circle
- vertical
- vertically

geometrisches Objekt
- Kreis
- Würfel
- Quader
- Zylinder
- Vollkreis
- Parallelogramm
- Rechteck
- Halbkreis
- Kugel
- Spirale
- Quadrat

Geometrie
- Winkel
- Bogenlänge
- Fläche
- Umfang
- Zirkel
- Zirkel
- verschieben
- verschoben
- Verschiebungsvektor
- Kantenlänge
- geometrisches Objekt
- Schnittfläche, Schnittpunkt
- Bogenlänge
- Zirkel
- parallel
- Umfang
- senkrecht [adj.], senkrecht [adv.]
- senkrecht [adv.]
- Radius
- halbkreisförmig
- verschieben
- verschoben
- räumlich
- Oberfläche
- tangential [adj.]
- tangential [adv.]
- Trigonometrie
- Einheitskreis
- senkrecht [adj.]
- senkrecht [adv.]

Mathematics / Mathematik

logarithm
   logarithm of a to base b

number
   complex number
   integer
   natural number
   rational number
   → real number

power
   base (of a power)
   exponent
   exponential function
   exponentiation
   x to the nth power
   x to the power of n
   power of ten
   nth power of x

product
   factor
   multiplication
   to multiply
   times
   a times b

proportionality
   constant of proportionality
   inversely proportional
   inversely proportional to
   proportional
   proportionality constant
   reciprocally proportional
   reciprocally proportional to

quotient
   to divide
   a divided by b
   division
   reciprocal

real number
   decimal place
   decimal point
   digit

Logarithmus
   Logarithmus von a zur Basis b

Zahl
   komplexe Zahl
   ganze Zahl
   natürliche Zahl
   rationale Zahl
   reelle Zahl

Potenz
   Basis (einer Potenz)
   Exponent
   Exponentialfunktion
   Potenzieren
   x hoch n
   x hoch n
   Zehnerpotenz
   n-te Potenz von x

Produkt
   Faktor
   Multiplikation
   multiplizieren
   mal
   a mal b

Proportionalität
   Proportionalitätskonstante
   umgekehrt proportional
   umgekehrt proportional zu
   proportional
   Proportionalitätskonstante
   umgekehrt proportional
   umgekehrt proportional zu

Quotient
   dividieren
   a geteilt durch b
   Division
   Kehrwert

reelle Zahl
   Dezimalstelle
   Dezimalpunkt
   Ziffer

negative sign	negatives Vorzeichen
norm	Betrag
positive sign	positives Vorzeichen
sign	Vorzeichen

root / Wurzel

to extract a root	radizieren, eine Wurzel ziehen
index	Wurzelexponent
radicand	Radikand
nth root of x	n-te Wurzel von x
square root	Quadratwurzel

straight line / Gerade

constant function	konstante Funktion
gradient	Steigung
linear function	lineare Funktion
shallow	flach
slope	Steigung
steep	steil
straight line through zero	Nullpunktgerade
y-intercept	y-Achsenabschnitt

sum / Summe

to add	addieren
addend	Summand
addition	Addition
plus	plus
a plus b	a plus b

trigonometry / Trigonometrie

adjacent	Ankathete
arcus	arcus
cosine	Cosinus
cosinus	Cosinus
hypotenuse	Hypotenuse
opposite	Gegenkathete
right triangle [A. E.]	rechtwinkliges Dreieck
right-angled triangle [B. E.]	rechtwinkliges Dreieck
sine	Sinus
sinus	Sinus
tangent	Tangens
triangle	Dreieck

vector / Vektor

arrow	Pfeil
to decompose a vector	einen Vektor zerlegen

Mathematics / Mathematik

direction	Richtung
final point of a vector	Endpunkt eines Vektors
initial point of a vector	Anfangspunkt eines Vektors
length of a vector	Länge eines Vektors
magnitude of a vector	Betrag eines Vektors, Länge eines Vektors
norm of a vector	Betrag eines Vektors, Länge eines Vektors
null vector	Nullvektor
to point (in a certain direction)	zeigen (in eine bestimmte Richtung)
to re-scale (a vector)	skalieren (einen Vektor)
resolution of a vector	Vektorzerlegung
to resolve (a vector)	zerlegen (einen Vektor)
scalar multiplication	Skalarmultiplikation
to shorten a vector	einen Vektor stauchen
start of a vector	Anfangspunkt eines Vektors
to stretch a vector	strecken (einen Vektor)
tip of a vector	Endpunkt eines Vektors, Spitze eines Vektors
vector addition	Vektoraddition
vector analysis	Vektoranalysis
vector decomposition	Vektorzerlegung

Vom selben Autor

"Physics Formulas and Tables"
A formulary for use in immersive teaching

Klett und Balmer Verlag Zug (2011)

ISBN 978-3-264-83994-4

http://www.klett.ch/

Diese physikalische Formelsammlung ist entstanden für den Gebrauch im Bilingual- bzw. Immersionsunterricht an Sekundarschulen oder Gymnasien im deutschsprachigen Raum.
Größter Wert wurde auf die Verwendung der korrekten britisch-englischen Fachbegriffe gelegt.
Das Buch erweist sich ebenfalls als hilfreich für Studenten und Lehrer an Universitäten und Hochschulen.
Zudem ist es von Interesse für Schüler, Studenten und Lehrer im englischsprachigen Raum.

www.ingramcontent.com/pod-product-compliance
Lightning Source LLC
Chambersburg PA
CBHW082206220526
45470CB00010B/3068